Disability and th

Rob Imrie has published widely in international journals on subjects spanning urban policy to industrial and economic planning. He has co-authored a book entitled *Transforming Buyer-Supplier Relations* (Macmillan, 1992) and is co-editor of *British Urban Policy and the Urban Development Corporations* (Paul Chapman, 1993). His background is in geography, sociology and planning, and he has a chartered Town Planning degree and a doctorate in industrial sociology. He previously taught planning studies in the Department of City and Regional Planning, University of Wales, Cardiff (1985-91); at present, he is a lecturer in geography at Royal Holloway, University of London, teaching courses in economic geography, regional policy and local planning. He is currently co-ordinating an Economic and Social Research Council project which is investigating disability and access in the United Kingdom.

Disability and the City
International Perspectives

Rob Imrie

Paul Chapman
Publishing Ltd

To Sarah and Marian

Paul Chapman Publishing Ltd
144 Liverpool Road
London
N1 1LA

British Library Cataloguing in Publication Data
Imrie, Rob
Disability and the city
1. Architecture and the physically handicapped
2. City planning – Social aspects
I. Title
720.4'2
ISBN 1–85396–273–2

Typeset by Whitelaw & Palmer Ltd, Glasgow
Printed and bound by The Baskerville Press, Salisbury

A B C D E F G H   9 8 7 6

Contents

Acknowledgements

I would like to thank Sue Lambert for permission to reproduce figures 4.2 and 4.3 from her book *Form Follows Function*. Unfortunately, despite efforts to do so, it was impossible to trace the copyright for figure 4.1. I would also like to acknowledge Stephen Ward for his permission to reproduce figure 6.1. All other illustrations are from the personal collection of the author. I am grateful to Pion Ltd for giving me permission to reproduce sections of the article Imrie, R. and Wells, P., 1993, Disablism, Planning, and the Built Environment, *Environment and Planning C: Government and Policy*, Vol. 11, Ch. 2, pp. 213-31. Thanks also to my colleague Peter Wells for his permission to reproduce parts of this article. Chapter 3 is substantially based on the article Imrie, R., 1996, Equity, Social Justice, and Planning for Access and Disabled People: An International Perspective, *International Planning Studies*. I thank the publishers, Carfax, for granting me copyright clearance to reproduce sections of the article in this book.

The author and publishers are grateful to all who gave their permission for the use of copyright material. They apologize if they have inadvertently failed to acknowledge any copyright holder and will be glad to correct any omissions that are drawn to their attention in future reprints or editions.

Preface

The long term physically impaired are neither sick nor well, neither dead nor well, neither dead nor alive, neither out of society nor wholly in it. They are human beings but their bodies are warped and malfunctioning, leaving their full humanity in doubt. They are not ill, for illness is transitional to either death or recovery . . . The sick person lives in a state of social suspension until he or she gets better. The disabled (sic) spend a lifetime in a similar suspended state. They are neither fit nor foul; they exist in a partial isolation from society as undefined, ambiguous people.

(Murphy, 1987, p. 112).

People with disabilities are one of the poorest groups in western societies, and they remain estranged from involvement in the socio-institutional fabric of society. In particular, they lack power, education and opportunities, while they remain subject to forms of state control which continue to reinforce the pejorative image of disability as an abnormality. For most disabled people, their daily reality is dependence on a carer, while trying to survive on state welfare payments. Moreover, the dominant societal stereotype of disability as a 'pitiful' and 'tragic' state reinforces the view that people with disabilities are somehow 'less than human' (Scheer and Groce, 1988). In taking exception to these, and other related, conceptions of disability, this book explores one of the crucial contexts within which the marginal status of disabled people is experienced, that is, the interrelationships between disability, physical access and the built environment. It seeks to explore some of the critical socio-cultural and political processes underpinning the social construction and (re) production of disability as a state of marginalization and oppression in the built environment. In turn, such concerns are interwoven with a discussion of the changing role of the state in defining, categorizing, and (re) producing 'states of disablement' for people with disabilities.

Focusing primarily on the United Kingdom, although with a substantial discussion of disability and access issues in the United States of America, the book also considers the role of the 'design professionals', architects, planners, and building control officers, in the construction of specific spaces and places which, literally, lock people with disabilities 'out'. In particular, western cities are characterized by a 'design apartheid' where building form and design are inscribed with the values of a society which seeks to project and prioritize the dominant values of the 'able-bodied'. From the shattered paving stones along the high street, to the absence of induction loops in a civic building, people with disabilities face the daily hurdles of negotiating their way through hostile environments which the majority of us take for granted. Using a range of empirical material, the book documents how the environmental planning system in the United Kingdom is attempting to address and overturn the inaccessible nature of the built environment for people with disabilities, while discussing how disabled people are contesting and challenging the constraints placed upon their mobility and abilities to gain access to the cities.

In writing the text, I have assumed that readers have some knowledge of the British planning system. For those who have little or no understanding of the system, the texts by Cullingworth and Nadin (1995) and Rydin (1993) can provide some basic background. A number of people have given me invaluable help in the preparation of this manuscript and throughout the course of its writing. I would like to thank Huw Thomas for his support over the years and for the original idea of looking at access issues in planning. I have always found him to be a wonderful support both intellectually and otherwise. Peter Wells was also instrumental in getting me started on the idea of looking at disability and access issues in the built environment and we spent many a happy time trekking across South Wales interviewing access officers and planners in the early 1990s. I would also like to thank the wide range of individuals who gave up their time to talk with me about access and disability issues, including Jo Nichols, Ken Matheson, Richard Kember, Hassan, Geoff Dougan, Richard Skaff, Peter Myette, Harrie, Lois Thiebault, John Dobinson, Andrew Walker, and others too numerous to mention. Support has also been provided by the Centre for Accessible Environments in London and the Centre for Independent Living in Berkeley, California, while the ESRC has generously continued to support my research into disability and access.

My thanks are also extended to the University of London Central Research Fund for the research grant which provided me with the opportunity to visit the USA in March/April 1994. This allowed me to investigate the impacts and implications of the most far reaching piece of legislation to have been passed anywhere relating to the civil rights of people with disabilities, the Americans with Disabilities Act (1990). I would also like to thank Royal Holloway, University of London, for the three month sabbatical which permitted me to draft some of the earlier chapters. Part of this period, in early 1995, was spent in the Department of Geography at the University of Waterloo in Canada and I would like to thank Professor Ron Bullock for providing me with office space and a fully functional word processor. This was a productive period and I would especially like to thank Tod Rutherford for his spare room, intellectual support and friendship, while I will always be grateful to him for introducing me to Heather, Lara, and Patrick. Finally, I would like to acknowledge the friendship and support of Sarah and Marian who both read drafts of the manuscript and continue to give me both intellectual and emotional support.

1

Disability and the City: an Overview

INTRODUCTION

The reasonable man (sic) *adapts himself to the world, the unreasonable man persists in trying to adapt the world to himself. Therefore, all progress depends on the unreasonable man.*

George Bernard Shaw

In the last decade access for disabled people to public buildings and facilities in the cities has become an important part of the political agenda, and many public authorities internationally are promoting strategies for 'accessible built environments' (Leach, 1989, Millerick and Bate, 1991). In particular, there is more awareness that disabled people, in their everyday lives, are having to confront hostile built environments, ones where access to buildings, streets, and places, is often impossible (Barnes, 1991, Imrie and Wells, 1992). There is also a recognition of the marginalization of disabled people in society, and the perpetuation of a segregationist ethos which not only continues to place disabled people in special institutions but also reinforces a number of problematical, pejorative and depersonalizing stereotypes (e.g. cripple, spastic, the handicapped, etc.). As Liachowitz (1988) suggests, such views continue to set the disabled apart from the (so called) able-bodied and nowhere more so than in the built environment which, as Hahn (1986a) concludes, is characterized by a thoughtless lack of design and planning in public and private buildings.

For instance, day-to-day artefacts, which the able-bodied take for granted, are usually (literally) out of reach, or unavailable, for the wheelchair-bound person (Barnes, 1991). For example, most cash dispensing machines are placed too high for wheelchair users to reach, while clothes retailers have few (if any) changing facilities for people in wheelchairs. Moreover, Barnes (1991) indicates that society's ignorance of sign languages generally excludes the deaf and hard-of-hearing from a range of public places, and concludes that disabled people's ability 'to perform even the most routine of daily tasks is thus severely diminished because of a predominantly inaccessible environment' (p. 180). Indeed, the urban environment is generally inaccessible for a range of people with disabilities, characterized by, for example, the interwar expansion of the suburbs, which, aligned to the postwar spatial divisions of city functions, generated cities which increasingly placed a premium on individual mobility. As Hahn (1986a) notes, in her discussion of Los Angeles, for people with disabilities the city is a vast desert containing few oases.

Such forms of marginalization, disadvantage, and oppression facing disabled people has led Leach (1989), amongst others, to recognize that disability is as much an equal opportunities issue as race and gender. As Hahn (1986a) suggests, in terms of ease or comfort most cities have been designed not merely for the non disabled but for a physical ideal that few humans can ever hope to approximate. Indeed, recent debates on access and mobility for the disabled have coined the term 'disablism' to refer to ideologies and values, which legitimize oppressive practices against disabled people, purely on the basis that they have a physical and/or mental impairment. In particular, disablist conceptions are infused with a range of problematical assumptions including the notion that disability is reducible to functional, physiological limitations, the assertion of the normality (and naturalness) of ablebodiedness, or what one can term the propagation of ableist socio-cultural values. In turn, the assertion of ableist values is intertwined with the notion that disability is abnormal, even a product of deviant behaviour, and it is assumed that the goal of society is to return disabled people back to a normal (able-bodied) state (whatever that is). Any sense of celebrating, even recognizing, the vitality of difference seems beyond the disablist nature of socio-institutional values and practices.

This chapter outlines the interrelationships between the changing structure of the modern city and the development of a built environment which is disablist, actively discriminating against the physical mobility and access needs of a significant proportion of disabled people. Part of the argument will reflect Davis' (1985) assertion that the spatial structure of the modern city reproduces dominant power relations, so contributing in subtle ways to the oppression and exclusion of large sections of the population. Such perspectives have been especially well developed by particular strands of feminism where the built environment is conceived as 'gendered' or, as Wajcman (1991) notes, 'sexual divisions are literally built into houses and indeed the whole structure of the urban system' (p. 110). In particular, I draw on the influential work of Iris Young (1990) to argue that people with disabilities are locked into systemic structural relations of oppression and domination constitutive of societal-wide values and practices implicated in the (re) production of the marginal status of disabled people. I utilize her framework of social oppression to provide an overview of the contrasting forms of discrimination and oppression which disabled people have to face in different spheres of their social and economic lives, while introducing the reader to the specific issue of mobility, access, and marginalization, as one subset of problems facing disabled people. In doing so, I seek to develop Barnes' (1991) view that institutional discrimination and oppression against disabled people are never more obvious than the restrictions placed on physical mobility and access by a poorly designed built environment. The chapter will conclude by outlining the themes of the book and its overall structure.

DISABILITY, OPPRESSION, AND DOMINATION: THE CRITICAL DIMENSIONS

Academic and policy communities continue to pay scant attention to disability and the experiences of disabled people (for exceptions, see Barnes, 1991; Dalley, 1991; Oliver, 1990; Townsend, 1979). Indeed, some research indicates the weak and marginal position of disabled people in society, yet, as Oliver (1990) notes, academics in the social sciences have researched and written little on their experiences (although, for exceptions, see Johnson, 1988 and Morris, 1991, 1992). For instance, as Townsend (1979) notes, even when employed, disabled people are more likely to be low paid and even less likely to have assets in the form of savings, personal possessions, and consumer durables. Similarly, Harris *et al*'s (1971) national survey in the UK found high rates of unemployment among disabled people, while a recent survey concluded that 25 per cent of disabled people are registered as unemployed, double the UK national average (Dalley, 1991). Moreover, an Office of Population Censuses and Surveys (OPCS, 1987) report notes that disabled adults in the UK tend to have incomes similar to those of pensioners while incurring much greater expenditures (to pay for the provision of disabled aids, etc.).

The relative deprivation of disabled people is particularly significant with regard to mobility and access (Gilderbloom and Rosentraub, 1990). As an OPCS survey (1971) in the early 1970s revealed, over 25 per cent of disabled people in the UK were living in 'poor' conditions, usually without an accessible indoor toilet, while most disabled people were living in ordinary dwellings with no special design features or facilities. The survey concluded that one million disabled people living at home were in need of rehousing or required substantial improvements to their existing dwellings. Moreover, despite calls to end the segregation of disabled from able-bodied people, a wide range of evidence concurs that little has changed (Garety and Toms, 1990; Reisen, 1990). For instance, as Barnes (1991) notes, only one in five disabled children attend 'normal' schools in England and Wales, while highlighting the 'prisoner syndrome' with every two out of five disabled people dependent on someone's support to enable them to take trips out of their homes for shopping, recreation, and other pursuits.[1]

Such patterns of both institutionalized dependence and estrangement are, for Young (1990), a key indication of the weak and socially unjust position of disabled people in western societies. Indeed, as Young suggests, it seems clear that a range of groups, including ethnic minorities, gay men, lesbians, women and, of course, people with disabilities, experience forms of social and institutional injustice on a daily basis which are antithetical to a just society (Harvey, 1973; Young, 1990). The manifestations are wide-ranging and, while their precise forms are contingent on a range of specific socio-spatial and temporal conditions, the commonality of experiences, of what Young terms the 'marginals', include their systematic exclusion from positions of socio-institutional and political power, a societal stereotyping of them and their lifestyles, where they are often presented as 'deviant', and a societal-wide conception, indeed, insistence, of 'their differences' as having little or no socio-cultural relevance and/or significance (Barnes, 1991; Morris, 1992; Oliver, 1990; Rose, 1990).

While injustice is traditionally conceived of as a distributional phenomenon, of the structural forms of who has access to, and use of, material goods, Young argues that such notions need to be extended to consider 'institutional issues of decision making structures and culture' (p. 24). In noting this, Young is echoing Rawls (1971) who, as she acknowledges, conceives of justice as including the 'rights and duties relating to decision making, social position, and power, as well as wealth or income' (p. 24). In this sense, Rawls, Young and others have a concern with conceiving justice (or its corollary, injustice) as a series of social processes rather than in terms of some end pattern or state. Thus, as Young argues, 'being enabled or constrained refers more directly, however, to the rules and practices that govern one's actions, the way other people treat one in a context of specific social relations, and the broader structural possibilities produced by the confluence of a multitude of actions and practices' (p. 26). For people with disabilities, for instance, their opportunities to work or gain access to a building are locked into systemic forms of such broader structural relations, to the prejudices held by society, and to their systematic exclusion, indeed segregation, from the socio-institutional mainstream.

As Young notes, injustice should be conceived of as oppression and domination, where the former is defined as 'systematic institutional processes that prevent some people from developing and exercising their capacities and expressing their experiences' (p. 71). As Oliver (1990) argues, and as subsequent chapters will show, few people with disabilities, for instance, occupy positions of power and, as Young (1990) comments, they are generally inhibited (prevented) from expressing their feelings and perspectives on social life where others are able to listen. People with mental impairments, for instance, are often ignored because mainstream society has little time or patience to try to understand the 'non standard' ways in which they try to communicate, while it is well documented that some care institutions do little more than to impose authoritarian regimes on such individuals (Morris, 1992). As Nichols (1995) notes, in discussing her time as a care worker:

for five years I worked in a care institution for people with mental disabilities. They had to conform to the regime, taking baths once a week at a set time . . . if they didn't like the time they didn't get a bath for that week . . . with those that couldn't communicate in the normal way, well they were just pushed out of the way and forgotten about . . . they obviously had needs but just couldn't express them in the obvious way, but it's wrong to just then impose things on them that they may not want.

Nichols' testimony is more common than not and, as Finkelstein (1993) and others have documented, while resource constraints in the last few years have, in part, underpinned the nature of care practices, such (oppressive) ways of attending to many people with mental impairments seem to be more for the (cost) benefits of the institutions and, so some would say, for the convenience of the staff, than for the care needs of the clients who live within them (see French, 1993).

In turn, such structures (re) produce relations of domination which Young (1990) argues are the core experiences of the day-to-day existence of many people with disabilities. For Young domination refers to the institutional conditions that inhibit people from participating in determining their actions or the conditions of their actions. As she notes, people live within such structures of domination 'if other persons or groups can determine without reciprocation the conditions of their action' (p. 38). And, for disabled people, there is widespread documentation of such institutional non reciprocity, from what Morris (1993) terms the benevolent paternalism of the voluntary sector residential homes of the 1950s in the UK, to the establishment of rehabilitation units in postwar USA, places which (re) produced cultures of care and dependence. As Oliver (1990) notes, commenting on the emergence of care services, 'despite the affluence of the postwar years . . . it soon became clear that disabled groups, among other groups, were not having all their needs met and, often, even those that were acknowledged, were being met in inappropriate or oppressive ways' (p. 114).

For Young, then, oppression is injustice and justice, first and foremost, refers to the institutional conditions necessary for the development and exercise of individual capacities and collective communication and co-operation. Indeed, while oppression, as injustice, is traditionally conceived of as a form of tyranny and/or colonial conquest, for Young, it can also refer to the 'everyday practices of a well intentioned liberal society' (p. 41). As Young lucidly argues:

> oppression in this sense is structural, rather than the result of the intentions of a tyrant. Its causes are embedded in unquestioned norms, habits, and symbols, in the assumptions underlying institutional rules and the collective consequences of following those rules . . . it refers to the vast and deep injustices some groups suffer as a consequence of often unconscious assumptions and reactions of well meaning people in ordinary interactions, media, and cultural stereotypes, and structural features of bureaucratic hierarchies and market mechanisms – in short, the normal processes of everyday life (p. 41).

In developing this, Young refers to five variants of oppression, multiple and/or interconnected forms which singularly or collectively define contexts within which particular groups, such as people with disabilities, experience their disaffection from the mainstream of society: exploitation, marginalization, powerlessness, cultural imperialism, and violence. As Young rightly argues, these five forms are neither mutually exclusive nor are they exhaustive in defining and delimiting societal forms of oppression. However, they do represent the dominant axis along and through which particular individuals and groups experience the multiplicity of exclusions which render them more or less invisible. Yet, one must recognize that Young's multiple forms of oppression represent something of an ideal type and are limited by implying a universal experience of disability. Indeed, while the categories tend to define social processes in a broad, generalized, way, it is clear that people with disabilities, for example, experience urban living in a variety of complex forms which cannot be necessarily captured by such categorizations. Nonetheless, as a heuristic device, the categorizations are a useful starting point for situating the lived experiences of disabled people in some form of non essentialist structural context.

Young's first category, *exploitation*, is closely aligned to the neo-marxian idea that oppression occurs through transferring the results of the labour of one social group to benefit another. A common feature of experiences of people with disabilities, is their exclusion from the labour market, but where inclusions do occur they tend to consign disabled people to low paid, low waged, unskilled occupations. The evidence for this has been well documented in that people with disabilities, where and when they work in employment, occupy menial, often service, occupations – jobs which are lacking in autonomy and, as Young describes them, 'auxiliary work, instrumental to the work of others, where those others receive primary recognition for doing the job' (p. 52). Their remuneration for work rendered is poor and government research for people with disabilities in work in the UK shows that they are locked into unskilled manual occupations receiving only two thirds of the salary of their 'able-bodied' counterparts (OPCS, 1987, 1989). Not surprisingly, people with disabilities are generally excluded from professional occupations and, as figures released by the Equal Opportunities Division (1993) within the UK Civil Service show, less than one per cent of positions at grade 7 or above across the whole of the civil service were occupied by disabled people.

Such status is linked to Young's second category of oppression, the *marginalization* of disabled people, and, as Young argues, the classical (marxian) definition of 'marginals' are people whom the system of labour cannot or will not use. In particular, disabled people represent a whole category of persons who have been more or less expelled from participation in many spheres of social life and, as Gooding (1994) recounts, while the official unemployment rate in Britain in 1994 was 9 per cent, for people with disabilities it was 20 per cent. As Gooding (1994) notes, a major underpinning of discrimination in employment relates to the persistence and perpetuation of the idea that disabled people are characterized by poor and/or limited work abilities and are prone to high levels of sickness and absenteeism (Kettle,

1979). Yet, as some research has documented, such notions are largely mythical, with disabled people, in terms of productivity and absenteeism, being little different from anyone else (Kettle, 1979, US Department of Labour, 1948). However, such stereotypical conceptions of people with disabilities are hard to dispel and a recent survey of UK employers by Workability (1990) indicated that 40 per cent would not employ a disabled person under any circumstances.

While material deprivation, as a facet of socio-economic marginalization, is a crucial aspect of the status of people with disabilities, as significant are other interdependent forms of marginalization (as forms of oppression). One relates to the dependence of disabled people on the welfare state and the production of injustices 'by depriving those dependent on it of rights and freedoms that others have' (Young, 1990, p. 54). As Morris (1993) notes, only 30 per cent of disabled people of working age in western Europe have a job and the social security system, therefore, is the primary mechanism in determining the standards of living which the majority of people with disabilities have. However, as Morris and others, have documented, the basis of the system is one which reinforces dependency while penalizing autonomy in that disabled people have to emphasize their incapacities (rather than capabilities) to qualify for benefits (also see Barnes, 1991; Swain et al, 1993). Thus, the recent changes to the UK system of 'incapacity benefits' for people with ambulant impairments require them to demonstrate to the Department of Social Security their capability or not in achieving measurable tasks of 'ability', like being able to walk unaided for a prescribed distance.

In particular, the dependence of disabled people on welfare services, while not oppressive in itself, has, in western societies, generated socio-political relations between professionals and clients which, as chapters 2 and 3 will highlight, tend to subject the latter to arbitrary, even punitive, and patronizing policy regimes. Thus, the oppressive and iniquitous relations within which people with disabilities have to lead their lives is being compounded by the increasing emphasis that social and welfare policy is placing on individuals 'to help themselves'. The USA provides some apt illustrations where, despite years of campaigning for integration, and the propagation of mainstreaming policies, people with disabilities are still seen as 'requiring treatment' and special attention. Thus, as Dougan (1994) recounts, builders still construct places for segregation and in Washington D.C. 'we've got a number of estates that lump all the physically disabled and elderly in set-aside places, what we locally call the GIMP ghettoes'. For Young (1990), such illustrations indicate the possibilities for producing forms of dependence which are unjust or, as she concludes, 'dependency in our society thus implies, as it has in all liberal societies, a sufficient warrant to suspend basic rights, to privacy, respect, and individual choice' (p. 75).

This is especially evident in one of the crucial aspects of disabled people's lives, the educational system. As Tomlinson and Colquhoun (1995) have documented, the further education curriculum in the UK emphasizes skills and the acquisition of 'core competencies' and, as they comment, 'those with special needs and/or disabilities . . . are urged that the way to find employment is by constant investment in the self by acquisition of skills and competencies' (p. 193). Thus, contemporary skills training in the UK is being driven by the National Vocational Qualifications (NVQs) yet the basis of such qualifications is itself predicated on maintaining distinctions between those who can 'generate valid evidence of competence' and those who, in being unable to do so, are deemed unfit for work. As the National Council for Vocational Qualifications has intimated, 'there is no question of the standards being adapted in such a way that they lose their relevance to employment, that would be in nobody's interest' (in Hillier, 1993, p. 2). This, then, brings to the fore the myth of meritocracy, that the individual investing in themselves and building up their stock of human capital, in itself, guarantees a position within formal labour markets (see Rees *et al*, 1992).

The myth of meritocracy is, however, premised on western conceptions of justice which hold that (work-related) positions should be awarded to the most qualified individual, that it is possible to be fair by allocating, for example, jobs in a value-neutral way to those with the appropriate, technical, skills which match the job vacancy. Yet, as many have argued, it is more or less impossible to separate values and cultural conditions from the decisions which are made about who should be allocated particular jobs and/or goods or services. As Young (1990) argues, the notion that somehow a technical neutrality is operating to produce the hierarchical division of labour is a convenience because it then legitimizes the persistence of patterns which see people with disabilities at the margins of the labour market ('they are there because they haven't acquired the skills, not because of societal and/or institutional prejudices'). As the ideology of technical neutrality presents itself, how could it be otherwise, given the person's disability and/or inabilities to perform the particular tasks which are required from the work organization. Yet such assumptions are problematical because they fail to consider the discriminatory and prejudicial practices of institutions and the frequent exclusion of disabled people because of the way in which institutional policies and practices favour the so-called able-bodied.

Such suspensions of basic rights are linked to a third categorization of the oppressed status of people with disabilities, that of their *powerlessness*. The powerlessness of disabled people is, in part, a function of their societal marginalization and of their weak and dependent relationship with the formal labour market. As Young (1990) notes, powerlessness is linked to positions within the division of labour and most people with disabilities have few formal work skills, no real technical expertise, and few possibilities for exercising decision-making power over their day-to-day experiences. In particular, the status of disabled people is such that they are daily exposed to disrespectful treatment. Indeed, the powerlessness of people with disabilities has been substantially reinforced by the alienation of disabled people from influencing social policy agendas in any ways which depart from the strictures laid down by professional welfare workers. As Barton (1993) notes, 'the overwhelming activities and noises with regard to both policy and practice have emanated from professionals. It is professional values and objectives which have defined need and practice' (p. 235) (also see chapter 3).

The powerlessness of disabled people does not exist independently of specific normative ideologies and values of what constitutes disability (and ablebodiment) and Young (1990) uses the term *'cultural imperialism'* to denote how oppressed groups 'experience how the dominant meanings of society render the particular perspectives of one's group irrelevant at the same time as they stereotype one's group and mark it out as the Other' (p. 59). As chapters 2 and 3 will explore in some detail, the substance of culturally imperialist values in society revolves around all of the daily contexts of disabled people's environments, from the shop with no disabled person's toilet, to the segregated special schools or units which label people with disabilities as somehow abnormal. For Young (1990), the dominance of 'able-bodied' values is wholly infused in all of our cultural practices and values, indeed, in the very substance of our languages. It is also reinforced by what Nelson (1994) refers to as a 'legacy of negativism' by society towards disabled people, a general treatment which is 'degrading and dehumanizing'. As Nelson recounts, various studies show how media, for example television, stigmatize disabled people by presenting them as 'not quite normal'. Thus, Leonard's (1978) study of television portrayals of people with disabilities concluded that the traits most depicted were 'dull, impotent, selfish, defensive, and uncultured' (p. 350).

The negative association of disability, then, represents a form of oppression which, as Morris (1993) notes, is underpinned by the possibilities for intimidation and ridicule. Indeed, a key form of oppression which people with disabilities have to confront is verbal and/or *physical violence*, usually in the form of acts of humiliation or stigma (see chapter 2 for an expansion on these themes). As Young (1990) notes, cultural imperialism intersects with violence and

> the culturally imperialized may reject the dominant meanings and attempt to assert their own subjectivity, or the fact of their cultural difference may put the lie to the dominant culture's implicit claim to universality. The dissonance generated by such a challenge to the hegemonic cultural meanings can also be a source of irrational violence (p. 63).

Xenophobic violence against women and ethnic minorities is well documented, yet little has been written about the relationships between violence and disability (Wolf, 1990). What documentation does exist tends to suggest that violent acts against disabled people occur primarily through forms of verbal and mental abuse, and/or physical separation (segregation) from the mainstream. As Young notes, the daily reality for many oppressed groups is direct intimidation but also the knowledge that they are liable to violation precisely because of their disability. Violence has also been institutionalized against people with disabilities and the professionalization of medicine in the nineteenth century reinforced the notion that disabled people were 'diseased' and required 'treating' (see chapter 2). This often involved incarceration and the use of the coercive violence of state restraint. Indeed, throughout the twentieth century, many people classified as 'mentally ill' have been subjected to the 'experiments of the day' and, as Wolf (1990) notes, 'electric shock is not just a metaphor' (p. 250).

Indeed, for many people with disabilities, part of the control exercised by state institutions, especially in the mental asylums of the nineteenth century, was the use of galvanic shocks. Showalter (1987), for example, compares the use of electroshock therapy on women asylum patients as resembling 'death and re-birth' or 'the trappings of a powerful religious ritual, conducted by a priestly masculine figure . . . [its magic] comes from its imitation of the death and rebirth ceremony. For the patient it represents a rite of passage in which the doctor kills off the bad crazy self, and resurrects the good self . . . a good self born again' (quoted in Wolf, 1990, p. 250). Indeed, people with disabilities have, at different times, experienced a societal xenophobia and the eugenics movement of the late nineteenth and early twentieth century emphasized the function of straining body types, of a genetic engineering, in order to realize the full potential of genetic endowment. The natural corollary of such violent thinking was that, for instance, during the second world war, Nazi doctors, were involved in sterilizing disabled people while their classification of people with disabilities as 'unfit' or as 'less than alive' was an important ideological prop in 'easing the doctors' conscience' (Wolf, 1990, p. 265).

One of the paradoxes with Young's conceptions is the way in which, on the one hand, they recognize the diversity of group experiences of oppression and domination yet, on the other hand, the tendency to elide the group experiences with those of individuals as though they are one and the same. Moreover, it is not wholly clear how the experiences of oppression and domination may vary by, for example, race, class, gender, or disability, or how such variety might be affected by the cross-cutting nature of such categorizations. Yet, as Bowe (1978) indicates, in the USA having a disability is, in part, linked to class status in that lower income people are twice as likely as middle or upper income groups to acquire a physical disability. Thus, in 1978, it was estimated that 20 per cent of all families on welfare in the USA, were in such a state because of the major breadwinner becoming disabled. There is, also, a tendency in Young's work to reduce the systemic structures of oppression and domination to the workings of the labour market and the division of labour in society, although this seems unidimensional and directional in terms of causal relationships.

Yet, the strength of Young's work is the way in which it questions what are often regarded as self evident, 'face-value' concepts by seeking to differentiate them and get behind the complexity of the substantive socio-institutional structures which, in large part, determine their form and content. Indeed, Young's insistence that experiences are multiple and differentiated is crucial for the understanding of disability, yet the term 'disability' tends to be chaotic because it suggests that there are a commonality of types and experiences which can be defined in and through the term 'disabled'. While not denying that people with disabilities experience some common forms of oppression, and encounter similar situations, the defining content of the term 'disability' revolves around the culturally imperialist values of an ableist society which seeks to reduce the essence of the body to 'types' (also, see chapter 2). In particular, 'disability' has, historically, been locked into a dualism with the notion of 'ability' and people have been consigned to one or the other. Yet, as Hall (1995) states, the problem with this is that the categories are presented as 'fixed states, unchanging and unaffected by social processes, clear images laden with social and cultural values' (p. 3).

Indeed, the concept of disability tends to be seen as naturalized with, as Hall argues, 'a person allocated to one side or the other on the basis of medical appearance and behaviour, and once allocated, to continue to look and behave in a clearly defined manner' (p. 3). For people with disabilities, such dualistic categories, of 'abled' and 'disabled' are powerful in reconstructing 'their' identities and of propagating stigma and reinforcing forms of social status. Moreover, such values tend to be conceived in ableist ways which reinforce the notion that a person with a disability is someone failing to measure up to, what Morris (1993) terms, the 'masculine values of strength, physical ability, and autonomy' (p. 87). Yet, as a range of authors has argued, such dualistic thinking tends to hide the complex and contested socio-cultural processes and practices which underpin the construction of dualities and, as Hall concludes, there is no easily defined state of being 'disabled' or 'ablebodied' in that such categories are constantly being challenged, contested, and transformed in wider socio-political and institutional practices (Birkenbach, 1993; Gleeson, 1995). Part of such processes relates to the interaction between 'states of disability' and geographical specificity, a theme I now turn to.

SOCIO-SPATIAL REPRODUCTION, DISABILITY, AND THE BUILT ENVIRONMENT

The dimensions of oppression and domination, outlined in the previous section, do not exist independently from socio-spatial processes, and geography, as this section will explore, is a key underpinning in the (re) production of particular forms of disablist social relations. Indeed, in extending some of the debates in the last section, this part of the chapter develops the thesis that the social construction of disability, attitudes towards it, and the development of disablism, are linked to the creation of the built environment, both those environments which seek to serve people with disabilities (e.g. the special schools) and those which effectively exclude them. As a range of commentators notes, the built environment has a physical inertia that resists change to the extent that many come to view city structures and spaces as almost fixed and immutable (see, for example, Savage and Warde, 1993). Such notions underpin, in part, the poverty of debate about the social construction of environments, and about how the city is dynamic and changing, and about the capacity of people to utilize skills and technologies to engender barrier-free places. In particular, it is popularly assumed that, because the built environment seems to facilitate access for the majority of the population (which is a contestable notion), it is the responsibility of the minority to cope by overcoming their handicaps and/or compensating for them.

Yet a plethora of research indicates that built environments are not naturalistic nor do they take shape independently of wider socio-political structures and processes (see, for example, the volume by Watson and Gibson, 1995). Rather, they are inscribed with and, in part, determined by wider socio-institutional relations. In particular, social inequalities in the city, as one of the more significant spatial materialities, are mapped by the complex spatial mosaic of place-based segregation, like suburbanization, ghettoization and gentrification, and, in this sense, such forms of segregation are expressive of the culturally imperialist values of society. Through the work of neo-marxist and neo-weberian, and feminist, scholars, a more sophisticated understanding of social inequalities and their spatial representations has emerged. Yet, as Savage and Warde (1993) have noted, the determinants of everyday life of subcultural groups, which emerge on the basis of material, socio-spatial and cultural inequalities, remain obscure. This is an apposite comment in the context of people with disabilities, for we know little about their interactions with the physical and/or built environment, although it is clear that they experience it in multiple and differentiated ways. For most, their experiences tend to be characterized by finite and demarcated spaces or forms of exclusion from particular spaces that are readily accessible to the 'able-bodied' (for example, see Golledge, 1991, 1993).

For Laws (1994a), society is constituted in space and, as she argues, 'space acts as both a container and shaper of social processes' (p. 21). In this sense, oppressive or any other social relations are constituted in and by space. In illustrating this, Laws refers to the geographies of oppression experienced by black Americans and, in quoting Hooks (1984), notes:

> As black Americans living in a small Kentucky town, the railroad tracks were a daily reminder of our oppression. Across those tracks were paved streets, stores we could not enter, restaurants we could not eat in . . . Across those tracks was a world we could work in as maids, as janitors, as prostitutes . . . We could enter that world but not live there. We always had to return to the margin, to cross the tracks, to shacks and abandoned houses on the edge of town (p. ix, quoted in Laws, 1994a, p. 21).

The delineation of oppressive spaces takes many shapes and forms over and beyond this. For instance, the marginalization of disabled people from the workplace often has little to do with their impairments but is more likely to be related to an inaccessible built environment (Barnes, 1991; Imrie and Wells, 1993). In this way, spatial exclusion is literally built into specific places.

Moreover, the re-location of indigenous Americans into reservations is an indication of how space has been utilized as a 'social marker' and as a mechanism by which the state seeks to control the activities of those deemed to be 'subordinate'. The state is also implicated in influencing the housing patterns of disabled people, both in terms of limited provision and also for many people with disabilities, their 'choice' of what to live in and where to live is often one of just taking what is offered (Morris, 1993).

In attempting to theorize such contrasting patterns of spatial interaction, a range of themes provides some relevant insights into the construction of space and of its recursive relationship with the emergent social structures of the city. At a general level, as Weisman (1992) notes, access to space is 'fundamentally related to social status and power and that changing the allocation of space is inherently related to changing society' (p. 6). While this conception is deterministic in tone, it is suggestive of the possibilities for conceptualizing space as constitutive of political relations, or of how the inscriptions of the built environment are an active shaper of human identity. Indeed, as Weisman argues, physical and social space reflect and rebound upon each other: there is an ongoing dialectical relationship between the two. Thus, for Weisman, specific aspects of the built environment both reflect and reinforce social exclusions and are infused with specific forms of class, race, and/or gender biases. An illustration of this is gentrification, where rising land values act as a mechanism to drive out lower income people from particular neighbourhoods. In turn, the combination of the economic with the changing socio-cultural and symbolic nature of particular spaces is also crucial and periodic episodes of 'white flight', for example, from incoming black ethnic groups, have been a characteristic of the shifting spatial terrain of distinctive groups in postwar American cities (see, for example, Little et al, 1988; Matrix, 1984).

However, while Weisman's approach is useful as a starting point, it tends to suffer from an underlying essentialism. This conceives of geography as a reflection, for instance, of dominant, ableist values, and, in turn, ableist socio-institutional policies and practices as being a reflection of geography. Yet, such conceptions eradicate any notion of the specificity of space, of the multiplicities of geography, and of the importance of the socio-spatial context of interactions in the determination of social processes. As King (1984) notes, the built environment is essentially a social and cultural product, or, as he comments, 'society produces its buildings and the buildings, although not producing society, help to maintain many of its forms' (p. 3). In this sense, a crucial theme in understanding the evolution of disablism is the connection between the built form and social and economic power or, as Laws (1994b) notes, how 'social or collective identities might be manipulated through built environments' (p. 1788), a theme which I explore in more detail in chapter 4. In concurring with Laws, however, it is clear that the social construction of disability, the meanings we ascribe to it and, consequently, the perpetuation of disablism, are linked to the imposition of culturally imperialist values in the creation of specific types of built environments (see chapter 2).

Yet much of the literature on the interrelationships between the built form and social marginalization has tended to be deterministic and premised on simplistic and problematical accounts of the interrelationships between people and the built environment. The classic example is the feminist analysis of the separate spheres which conceives of spaces as constructed around a gendered duality. On the one hand, there is the private world of the domestic realm, places inhabited by women who are engaged in caring for children while creating the social conditions for the reproduction of the (male) breadwinner. On the other hand, there are the public spaces, the world of formal employment, places largely inhabited by men. Women are, according to the analysis, locked into a private realm characterized by the oppressive spaces of a fragmented suburbia, places bereft of social facilities or of the means to transport women to the more public spheres of the city centre. This conception, then, conceives of geographic space as a reflection of patriarchal values and, in turn, the particular spaces of patriarchy as one of the oppressive mediums through which women's subordinate status is maintained and (re) produced (see, for example, Matrix, 1984)

The separate spheres debate, while illuminating the nature of socio-spatial inequalities in the built environment, is flawed. Fundamentally, the utilization of a 'duality', as a spatial metaphor, tends to simplify what are differentiated and fragmented socio-spatial relationships. Thus, at one level the analysis fails to differentiate between different types of men and women and assumes that power operates in one direction, from men against women. Moreover, there is the stereotypical conception that all women occupy similar spaces, that is, the patriarchal, familial setting, the stay-at-home person looking after the children. Yet, clearly, this is at odds with the diversity of family structure and types, and the reality for most women is of increasing incorporation into the so-called public spheres of formal employment. In this sense, not only is there an elision between the public and the private, there is little that is coherent about the categories 'men' and 'women'. In addition, there is a tendency for activity patterns to be reduced to specific gendered flows, constrained to particular parts of cities. So, for women, certain parts of any city are perceived as 'off-limits', such as, for instance, public bars, yet such a conception is seemingly insensitive to the possibilities of transgressions, even the degendering, of such spaces.

Moreover, the approach is underpinned by a crude determinism in theorizing the interrelationship between women and the built environment, generating research and writing which did little to identify the multiple and complex nature of social oppression that many women (and men) were, and are, subjected to. It reinforced a crude, fundamentalist, part of feminism which failed to identify the differentiated structural situation of women and of their capacity to act on, and adapt to, specific types of environments. Any idea of women having some form of autonomy and/or abilities to be transformative agents was either weakly developed or absent. There is also an underlying functionalism and essentialism associated with the separate spheres analysis in that behaviour and 'states of being' are read off as being naturally derived from a particular socio-spatial location. Thus, as Young (1990) notes, 'the separation of urban functions forces homemaking women into isolation and boredom' (p. 246). Yet the difficulty with this is the reduction of all 'homemakers' into an homogeneous category and the assumption that boredom is a consequence of a particular social status and place (i.e the suburbs).

The dualism of public and private spheres has, however, some resonance in relation to disablism in the city if only to highlight the hidden and partially private status of people with physical and mental impairments. Indeed, the advent of the special institution, of segregated spaces to deal with the 'peculiarities' of the disabled, precipitated spatial markers which somehow set them apart, socially estranged and outside the mainstream of society, effectively ghettoized. Such an inscription was especially evident in the context of the nineteenth century mental asylums, the propagation of particular moral spaces, or places that the dominant socio-institutional practices defined as the 'appropriate' spaces by which to avert the gazes of the disabled. These ranged from the special schools segregated in particular parts of the cities, to the asylums purpose-built to both control and 'hide away' those that society deemed 'mad' and 'uncontrollable'. For Schull (1984) and others the segregated 'solution' was critically related to spatially shifting 'the problem', or a socio-institutional response which was wholly underpinned by a decision making apparatus which can be characterized more or less as non reciprocal in giving people with disabilities little or no voice in the policy fora.

In particular, the infusion of geographical landscapes with particular ableist spatial markers is evident everywhere from the steps leading into government buildings to the absence of visual aids to enable people who are hard-of-hearing and/or deaf from moving around their environments (see Mitchell, 1995). Moreover, many disabled people have little option but to stay at home because the facilities to transport them around either do not exist and/or are difficult to schedule, or are prohibitive because a carer, who may not always be available, is required to accompany them. In addition, clear spatial signifiers generate hostile and exclusionary spaces which prevent many disabled people from using particular environments. In the USA, for instance, cities still erect implicit signs to disabled people telling them not to live in certain areas. As Burgdorf and Burgdorf (1975) have noted, cities like Chicago, until the end of the 1960s, erected signs which stated, 'No person who is diseased, maimed, mutilated or in any way deformed so as to be an unsightly or disgusting object or improper person to be allowed in or on the public ways or other public places in this city shall therein or thereupon expose himself to public view'. Apparently, such ordinances can still be found in cities like Colombus in Ohio, and Omaha in Nebraska.

Laws (1994b) also sees the construction of socio-institutional spaces in similar ways and, in writing about the institutionalization of the elderly in Toronto, notes how space is a constitutive part of how the elderly come to be defined and recognized, that it is in and through space that the multiple meanings of 'being elderly' are realized. As she argues, 'environments designed specifically for older people have played, and continue to play, an important ideological role in public discourses about the position and status of older people' (p. 1791). Thus, for instance, environments for the elderly tend to be 'set-aside', lumped together in ghettoized formation, from, for example, the sheltered (note the protective terminology) housing schemes in the UK, to the large suburban nursing homes in the USA, placed in quiet and leafy places (and, of course, pandering to the white, affluent, elderly). Such environments are signifiers of 'difference', of a special treatment, and of the undesirability for different (age) groups to mix. Ultimately, ageism, as an oppressive social relation, is being reinforced much in the way disablist social values over the years have underpinned segregated spaces for people with disabilities.

In theorizing the nature of disability and the city a crucial concern is to generate some understanding of the interactions between the body as a physio-logical and socio-cultural artefact and the built environment (a theme I extend in chapter 2). As Grosz (1992) has asked, what are the constitutive and mutually defining relationships between bodies and cities and how is the latter involved in the social production of the former? Such issues have been explored in some depth in relation to sexuality and space, of seeking to document the 'sexing' of places, of how landscapes are inscribed with sexual power and of how such power, in itself, is a recursive shaper of social spaces (Colomina, 1992; Pile and Thrift, 1995). Likewise, the shaping of ableist spaces is crucially linked to conceptions of the body. Thus, as Knox (1987), Sennett (1994) and others have documented, classical architectural theory, for instance, revolved around a conception of the body which was wholly embodied by the symmetry of an 'ablebodied' man. Buildings like the Parthenon in Athens were constructed with what Sennett (1994) refers to as a 'systematic imagery', a gender partiality but also a conception of health, vigour, and bodily rightness as portrayed in the classical context of able-bodiedness (see chapter 4).

Thus, as Sennett (1994) acknowledges, cities are more generally conceived of as being constitutive of social processes in which the interlinking between materialism, idealism, and the human desire to ascribe meaning, are powerful shapers of the human experiences of the built form. One illustration relates to the ableist reproduction of maps as particular spatial representations. Cartography tends to reinforce ableist images and values of the built environment and, as Vujakovic and Mathews (1994) have noted, 'if cartographers or planners take upon themselves to produce maps and guides for people with disabilities without first seeking to understand that group's images of the environment, they are likely to imbue the product with their own values and meanings rather than those of the potential user' (p. 373). In this sense, as Vujakovic and Mathews note, maps tend to reproduce authoritarian images, reinforcing and legitimizing the status quo. They provide an example of a map, or guide, produced by Stirling District Council for people with disabilities which they argue is a good news map proclaiming the facilities available in Stirling – but remains silent on barriers or constraints to mobility – therefore there is no indication of the town's steep topography or information on surface condition. The map also generalizes features . . . the road network is presented as uniform giving the impression of ease of mobility (p. 363).

In addition, the institutional practices of the state are a locus for the (re) production of particular built environments and it is clear that postwar planning policies, both in the UK and the USA, had the effect of encouraging a built form which was not only gendered but also ableist in structure and design (a theme I explore in greater detail in chapters 5 and 6). One of the better documented examples of the emergent gendered spaces is of the British new towns where planning principles emphasizing the separation of activities were encouraged. In the British context, research shows how the new towns were spatially dispersed, thus placing a high premium on mobility, yet lacked the requisite public transportation networks to facilitate access for those without a private motor vehicle (see Hall, 1984). For feminist geographers, such spaces reflected what they have referred to as the 'macho myth of metropolitan architecture' (Little, 1994, p. 59). Indeed, examples abound of how the built form is an exclusionary and exclusive realm and, as Little notes, the circulatory routes in cities, for example, reflect an economic efficiency primarily aimed at maximizing speed of movement rather than facilitating access for pedestrians.

Indeed, for Laws (1994b), such environments have been (re) constructed by virtue of the sexist nature of urban policy programmes. So too for disabled people where their knowledge has been denied in the construction of urban space. In particular, the development of planning theories and practices has tended to reflect the dominance of white, ablebodied, men, where the pursuit of enlightenment values, such as economic efficiency, has been paramount (a theme I explore in more detail in chapter 6). Such values, in turn, have underpinned an intellectual concern with the construction of divisible, functional spaces, and the emergence of a town planning ideology in postwar Britain, premised on the creation of the separation of 'non conforming' uses, like heavy industry and housing, played a part in the spatial decentralization of functions which has heightened the inaccessibility of the city to many disabled people. In turn, the city of the late twentieth century places a significant emphasis upon mobility as a means of gaining access to particular places and resources, yet, as I shall recount, such tendencies have served only to reinforce the disablist nature of the built environment.

MOBILITY, OPPRESSION AND ACCESS

As Blomley (1994) has argued, mobility and access are intrinsically geographical yet, as he asks, is it possible to conceive of a specific geography of rights in connection with access to particular spaces and places? Such issues have a place in liberal philosophical thinking and, as Blomley recounts, 'rights and entitlement attached to mobility have long had a hallowed place within the liberal pantheon and, as such, mobility is part of the democratic revolution' (p. 413). Indeed, in both the USA and Canada, mobility rights have been formally recognized and, as a range of authors has noted, differential forms of mobility are important components in the construction of geographical landscapes, being predicated on, and simultaneously reinforcing, social power and socio-spatial inequalities (Blomley, 1994; Laws,1994a, 1994b). Thus, as feminist geographers, for instance, have documented, the threat of urban violence is important in denying particular spaces to women, while recent rounds of public transportation privatization in the UK have demonstrably reduced the mobility choices and options of a range of individuals, usually those who are disproportionately less well off and unable to find alternative means of mobility.[2]

Mobility, then, is both an expression of social inequality and one of the socio-institutional mechanisms for reinforcing it, so playing an important part in determining whether or not particular categories of people can gain access or not to particular places. For people with disabilities, mobility, or often its absence, is a crucial context for reinforcing their subordinate and marginal status in society, and, in this sense, is one of the crucial contexts underpinning disabled people's experiences of oppression and domination. Indeed, a range of empirical research indicates that mobility concerns are of crucial importance to disabled people. For instance, the fourth annual Independent Living Consumer Survey (1991), held in the USA, shows that the most significant 'areas of difficulty' that faced the respondents were firstly, mobility, followed by public transportation, bathrooms (especially away from home), then steps and street kerbs (see Table 1.1). Other research concurs with such findings, and the Disabled People's Transport Advisory Committee (DPTAC, 1989) in the UK, for instance, estimated that 12 per cent of the population were unable to gain access to public transportation, with, for example, only 9 out of 800 National Express coaches accessible to people in wheelchairs in 1990.[3]

Hobbes, in conceiving of mobility in terms of the liberty of the human body, regarded constraints on mobility as oppressive. Yet, for many people with disabilities, such liberties are constrained, channelled, even denied. Indeed,

Table 1.1   Top ten 'areas of difficulty' confronting people with disabilities

Mobility
Public Transportation
Bathrooms (especially away from home)
Steps/street curbs
Funding/finance
Getting up from a sitting position
Fatigue
Frustration/feeling overwhelmed
Travelling
Having to depend on others

Source: Independent Living Consumer Survey, 1991

design apartheid, which serves to segregate and separate the disabled person from the mainstream, is evident everywhere and some examples are highly illuminating of both the thoughtlessness of the design professions and the wider structural constraints that they have to operate under. For example, Figure 1.1 is an illustration of the main entrance to the American Institute of Architects in Washington D.C., an organization which, while professing itself to be in the vanguard of challenging disablist architecture and design, is actively excluding people with disabilities from entering through the front door. In this instance, wheelchair users, along with deliveries, are invited to enter around the back through the loading bay, a situation which is both demeaning and richly symbolic of the residual status of people with disabilities!

Figure 1.1 The main entrance to the American Institute of Architects Building, Washington D.C., is illustrative of the thoughtless ways in which built environments are constructed. Steps prevent wheelchair users from exercising their rights to use the front door. The sign above the wheelchair symbol is apt in that many people with disabilities feel that they are rarely treated any better than baggage or deliveries.

Yet, while this at least provides a means for wheelchair-bound people to enter buildings independently, there are many more examples where design aesthetics, often coupled with thoughtlessness in the design process, effectively serve to exclude people with disabilities (see Gilderbloom and Rosentraub, 1990; Davies and Lifchez, 1987). Thus, Figure 1.2 shows the typical situation where steps into shops are generally used in the building process yet represent a completely unnecessary design feature. In such circumstances, visually impaired people and people in wheelchairs are, in part, dependent on others to gain access to the premises, while many shop doors, of the type depicted in Figure 1.2, are too narrow for people in wheelchairs to gain entry. Such architecture, then, goes against what Davies and Lifchez (1987) regard as a 'liberating' form or where the individual user is encouraged and facilitated by design rather than being hindered by it. Indeed, hindrance and discouragement are experiences that disabled people encounter on a regular basis yet ones which often remain transparent to those without the same, or similar, condition.

Figure 1.2   At the back of the store a sign says 'more choice', yet for many people with disabilities the store offers 'no choice' because of the step into the shop. It poses problems for people with prams and pushchairs too, while the inside of the store, like many shops in the UK and elsewhere, is littered with obstacles so making it difficult for visually impaired and/or blind people to negotiate their way around without falling over and injuring themselves.

A paradox is that such transparencies are often reflected by organizations which, while seeking to encourage mobility and access, do so in ways which draw attention to the disability. So, for instance, Figure 1.3 is an illustration of the Federal Deposit Insurance Corporation building in downtown Washington D.C. where, although a separate main entrance door is provided for people in wheelchairs to gain access to the building, the conditions of entry require people to 'please ring and wait for an officer'. This is an instance where a person with a disability is somehow being told that they are inferior and/or that there is something wrong with them (because they require help to get into the building).

Indeed, environments are littered with paternalistic signs which reinforce the idea of people with disabilities as somehow having 'childhood status' or, as Moore and Bloomer (1977) note, 'ramps at every turn and over emphatic graphics are only some of the elements that scream that this is a safe environment for the handicapped' (p. 7).

Figure 1.3 This represents something of a lock-out. People with disabilities are asked to ring the bell for assistance to get into the building. The author rang the doorbell on three occasions and nobody came to answer it. Why should anyone with a physical and/or mental impairment be directed to an entrance which requires them to be assisted? For many people with disabilities, the existence of separate spaces of this type is deeply symbolic of their marginalization and oppression.

In addition, as Figures 1.4 and 1.5 indicate, people with disabilities are regarded with some indifference where measures to facilitate mobility and access are often presented as an after-thought with little regard to integration and/or forging independence of mobility. Thus, in Figure 1.5, of a government exhibition building in Wales, some problematical features are evident. While it may have won a national architectural award it is unlikely to win anything for accessibility. Even though a ramp is provided for wheelchair users it is too steep, slippery when wet and forces the wheelchair-bound person to seek help. It offers no help either to the hard of hearing and/or people with visual impairments in the form of signage, tactile walkways, etc. In both instances we have examples of disablist design, or public spaces that relegate certain types of people with disabilities, predominantly the wheelchair-bound, to spaces in which they are rendered invisible or visually subordinate, and from which they are more or less excluded.

Figure 1.4 For visually impaired and/or blind people, the pathway to the visitor centre in Cardiff has little colour differentiation or tactile paving while the final entrance is a uniform grey which gives the impression of the steps all merging into one.

The disablist nature of mobility and access in the built environment, therefore, is fundamentally problematical because, first and foremost, it constitutes a 'denial of place' to those who, through no fault of their own, are penalized by oppressive socio-institutional structures and practices from exercising choices over how to utilize space. Hahn (1986a), in considering the ableist construction of Los Angeles, provides an apt comment, in that, 'for a sizeable proportion of the disabled population, Los Angeles remains unexplored territory, an unending stretch of impenetrable geography that may never be experienced because of artificial obstacles imprinted upon the land by decisionmakers' (p. 280). Yet, in that this type of restrictive mobility and access has the capacity to infringe a whole range of life opportunities, it seems clear that mobility and access rights should be placed on the political agendas of governments seeking to transcend disablist environments. As Blomley (1994) notes, casting mobility as a 'right' potentially 'draws morally valued questions of choice into the equation so confronting the universal-asserted values of society with the notion that some can choose, others cannot'. Indeed, for Blomley, a rights enframing is far reaching because it provides the means 'by which relations of subordination can be politicized' (p. 417).

Thus, the disabled person, for instance, outside the Federal Deposit Insurance Corporation building in Washington D.C., regarding their mobility

Figure 1.5   The visitor centre in Cardiff has attempted to provide wheelchair access yet the angle of slope exceeds the recommended gradient of 8 per cent (or 1 in 12). For most people in wheelchairs, self propulsion requires a slope of at most 6 per cent (or 1 in 14) and even this is often too steep for many people.

and access as an intrinsic right, may seek to question what it is that effectively keeps them out and/or makes it more difficult for them to use the building than for other categories of users. In this sense, mobility and access-based rights have some potential to shift the self-understanding of disablist environments from the individual impairment to the wider structural contexts within which such environments are constructed. As Laclau and Mouffe (1985, quoted in Blomley, 1994, p. 417) have noted:

> the denial of free movement to women or the disabled as the abrogation of a right, rather than the result of the operation of an indifferent and neutral market, or a lamentable burden to be borne by a disadvantaged (and hence naturally marginalized) individual changes the stakes. An appeal to mobility rights also offers something different that perhaps, now more than ever, is needed, ensuring the means by which the politics of space can be brought under scrutiny and contested.

CONCLUSIONS

This chapter has presented disability as an oppressive social relation. It exists in all western developed societies and their social practices and policies represent an arena of contest over the meanings and definitions of 'being disabled'. In particular, disability is a socio-politically defined and contested identity (or identities) and, following Laws (1994a, 1994b), it seems clear that the built environment is implicated in socially producing and reproducing the identities which surround it. This, then, is the wider theme of the book where I seek to investigate the interrelationships between disability, access, and the built environment. In particular, there are a range of broader themes which the book is concerned with. Foremost is the concern with the socio-cultural and political processes underpinning the social construction and (re) production of disability as a state of marginalization and oppression. Throughout the text, disability is reaffirmed as a socio-political and cultural category, neither fixed nor unchanging, yet one which is firmly understood only by reference to the specificities of the socio-spatial, and temporal, conditions of its existence.

In turn, such concerns are interwoven with a discussion of the changing role of the state in defining, categorizing, and (re) producing 'states of disablement' for people with disabilities. In particular, the identities of people with disabilities are closely intertwined with their relationships with the state, from the incarceration of the 'mentally ill' in the asylums of the 19th century, to the perpetuation of paternalistic welfare of the type which underpinned the postwar 'welfare settlement' in the United Kingdom. For many people with disabilities, their life experiences have been, and continue to be, defined in and through state institutions, and a key aspect of their continual oppression, as later chapters will reflect, is the dominance of a biomedical model of disability, or the conception of disability as a disease. The text also reflects on the differences which physical impairment makes to the ways in which people with disabilities have to lead their lives and considers the emergence of a 'politics of difference' as a way in which disabled people are trying to fight against the social pathological underpinnings of biomedical conceptions in order to achieve social equality without necessarily denying (as 'ablebodied' society would like them to) their physiological states.

These issues are explored in relation to disabled people's physical access to the built environment and a major focus of the book relates to the role of the 'design professionals', architects, planners, and building control officers, in the construction of specific spaces and places which, literally, lock disabled people out. In particular, western cities are characterized by a 'design apartheid' where building form and design are inscribed with the values of a society which seeks to project the dominant values of the 'able-bodied'. From the shattered paving stones along the high street, to the absence of induction loops in a civic building, people with disabilities face the daily hurdles of negotiating their way through environments which the majority of us take for granted. Yet, the ways in which such places 'get built' are not simply the result of the thoughtlessness of design professionals but reflect a much wider complexity of socio-cultural and political processes. In particular, the emergent political economy of the last decade or so has been one in which, for instance, the property development industry in the UK has gained some latitude to design and build environments free from what successive neo-conservative governments have characterized as the 'dead hand' of state controls.

Thus, state planning codes and practices relating to access for disabled people in the built environment, have tended to reflect an emergent market utility where developers' costs and concerns are increasingly a key determinant in what gets built. Moreover, state legislation towards access, especially in the UK, is weak and reactive with governments emphasizing the importance of the provision of accessible places only if a market demand is expressed for them. Indeed, statutes tend to emphasize 'codes of conduct', 'negotiations', and 'voluntary' compliance which tends to weaken the ability of professional groups, like planners, to insist on the provision of accessible places. Ironically, the weaknesses of the state's response to access have served to illustrate that people with disabilities are not passive consumers of the built environment, and the book seeks to show how disabled people are contesting and challenging the constraints placed upon their mobility and abilities to gain access to the cities. While the main geographical focus of the book is the UK, extensive comparative references are made with regard to the USA, especially in chapter 3.

NOTES

1. Since the passing of the 1993 Education Act in England and Wales, which requires children with disabilities to be integrated into 'mainstream' schools, this situation may be changing. Yet the evidence seems to indicate that, while mainstream schools are developing 'special units' on site, disabled children's needs are still being neglected often because of the high levels of demand they are exercising on a resource base which, in the education sector, has been progressively cut over the last fifteen years or so.

2. However, the responses have, in effect, heightened disabled people's feelings of marginalization and exclusion in as much that the policy 'solutions' have been based either on welfare provision and/or segregation. Thus, the principal state benefit to facilitate mobility for disabled people is claimable through the Disability Living Allowance, while other schemes are able to compensate for additional costs of travelling to work, etc. In addition, segregated schemes, like Dial-a-ride buses, have emerged, yet, as a range of research shows, such services are irregular, underfunded, inconvenient, and provide little or no independence for the user (Barnes, 1991).

3. Yet the contrast between different countries in terms of access policy and provisions is quite marked. For instance, by 1980, Germany had 118 fully adapted rail carriages in use on its network while in the UK there were none. Moreover, in the Netherlands, all buses provide adapted seats and additional handrails which are brightly coloured to help people with visual impairments, while services are door to door. In contrast, in Leicester, one of the more progressive cities in the UK with regards to disability rights, since 1982 only three buses have been equipped to enable

wheelchair users to gain access while a shuttle service has been provided on eleven routes where detours can be made for individual passengers. Inconsistencies abound, however, and in Belgium, where there is a national policy to promote the economic integration of disabled people, the decision was made to exclude wheelchair accessibility to the newly reconstructed metro system (Daunt, 1991).

As Gilderbloom and Rosentraub (1990) note, significant transportation and environmental barriers were preventing many individuals in Houston, Texas, from taking part in the socio-economic activities of the city. In a study of 1640 residents in Houston who had special housing and transportation needs, the authors noted that three out of every five disabled persons and elderly did not have pavements or kerb cuts between their homes and the local bus stops, while a greater proportion of people with disabilities in Houston (71 per cent) did not have a kerb cut by their nearest bus stop.

FURTHER READING

The reading material on the oppression of people with disabilities falls into a range of different types. On the one hand, there are the more journalistic, personal accounts, like Morris (1993); while, on the other hand, there are texts which seek to place the oppression and marginalization of people with disabilities into particular theoretical frameworks, like Oliver (1990). Another useful read is Barnes (1991, reprinted 1994), while the reader by Swain, et al (1993), is comprehensive. One of the best books on the theme of social oppression and marginalization is Young (1990).

2

Theorizing Disability and the Environment

INTRODUCTION

The marginal and oppressed status of people with disabilities is a significant feature of our societies yet, in comparison with the study of racial and sexual oppression, it has received little attention by theoreticians in the social sciences (although, for exceptions, see Blaxter, 1980; Borsay, 1986; Finkelstein, 1982; Gleeson, 1995; Oliver, 1984, 1990, 1992). As Oliver (1990) notes, attempts to understand the nature of disability have either tended to utilize inappropriate theoretical perspectives or have focused on the social oppression of disabled people by collecting what Oliver refers to as 'non reflexive and abstracted data' (p. 1). In particular, a characteristic of disciplines like geography and planning is their fixation with what Oliver terms an abstracted empiricism or the use of methodological strategies to measure 'the extent of the problems disabled people have to face' (p. 1). Yet, as Gleeson (1995) notes, such strategies have generally failed to do more than provide broad-based descriptions of different conditions of disability, while ignoring the wider socio-political structures, and determinants, of disabling states.

The origins of empiricist documentations of disability were firmly linked to governments seeking demographic information to enable them to exercise social and economic control over their populations (see Stone, 1984). Indeed, western societies have increasingly been concerned with demographic classifications and categorizations, and, as Oliver (1990) argues, the rise of capitalism was crucial in the labelling of individuals in terms of their capacity to function or not within the emergent labour markets. Terms like disabled, able-bodied, abnormal and normal were, and still are, a crucial means for categorizing people by physiological function and, by the mid nineteenth century, the notion of normality as somehow good, valuable and worthy was being counterposed to that of abnormality, the deviant, the worthless, the disabled. Such notions, aligned to the rise of forms of methodological individualism, or explaining the nature or essence of society purely in terms of facts about individuals, set the broader theoretical context within which the dominant theorizations of disability have developed and through which our understanding of disabled states has emerged.

In particular, a medical model, or theorization, of disability has dominated conceptions of disability in the social sciences (Harris *et al,* 1971; Hari, 1975). It conceives of disability as an individual, physiological, condition which can somehow be treated and cured. Oliver (1990) refers to it as the 'personal tragedy' theory of disability or that conception which sees disability as something which is wholly a problem of and for the afflicted individual. In turn, the resultant discourses of disability have tended to 'blame the victim' which, as Imrie (1996a) notes, portrays people with disabilities as 'inferior, dependent, and, by implication, of little or no value' (p. 3). Moreover, while aspects of the medicalization, or functional limitations, paradigm have been questioned and, some would argue, discredited, it still retains a powerful hold over elements of public policy towards disability (Hall, 1995; Imrie and Wells, 1993; Oliver, 1990). This is especially so with regard to access policies, particularly in the UK, where often the questionable assumption is made that because the built environment facilitates access for most people, then it should be possible for disabled people to adapt their behaviour to the environmental constraints that they encounter.

This conception underpins what Gleeson (1995) terms the dominant methodological approach of the medical theorization of disability, a form of positivistic behaviouralism in which disability is seen as an adaptable physiological condition. This methodology conceives of disability as somehow unnatural, more or less biologically produced, and where the problems which face disabled people are the result of their physical impairments independent of the wider socio-cultural, physical and political environments (Farber, 1968; Hellman, 1984). However, as this chapter will explore, critiques of such perspectives have emerged, trying to set impairments in their socio-cultural contexts, to understand them less as a physiological condition and more as a socially derived, and conditioned, state (see especially Oliver, 1990). These critiques have led to a plethora of approaches, from what one might term the minority model of social oppression, where disability is conceived of as a product of oppressive, majority population values and attitudes, to historical materialism, which seeks to explain disability as a work-related and determined category (Gleeson, 1995; Golledge, 1991, 1993).

Yet such theorizations of disability are beset by a range of disputes concerning issues of who is disabled and how knowledge should be produced about such people and the disabling social relations they are embedded within (see Imrie, 1996a; Lennon and Whitford, 1994; Oliver, 1992; Zarb, 1992). In particular, a significant problem with most theories is an underlying reductionism whereby people with disabilities are presented as an homogeneous group, as though all their interests and needs are similar (Golledge, 1993). More generally, disability rarely features in social theory, while post modern discussions of social differentiation seem to exclude accounts of disabled people. In particular, as Morris (1991) notes, such exclusions are evident in writings about ethnicity and feminism and, as Fine and Asch (1988a) have argued, the absence of women with disabilities from feminist writings, and from participation in women's groups, undermines the 'call from post modern feminists to recognize differences because nowhere do they recognize disabled people as a group whose voice needs to be heard' (p. 4).

In addition, a common feature of the competing discourses is, as Oliver (1992) and Zarb (1992) have noted, a knowledge denial or the absence of people with disabilities actively contributing to the generation, theorization, and construction of knowledges of themselves and of the structural conditions underpinning their daily experiences. Not surprisingly, the marginalization of disabled people extends to academia and government policy departments where research about disability is formally 'produced'. Indeed, as Oliver (1992) and Imrie (1996a) have argued, the social relations of research production, underlying the study of disability, are problematical because they perpetuate elitist structures which conceive of the researched as somehow subordinate (or inferior) to the researcher. Heron (1981) finds the divide between those who claim they know ('the experts') and those who do not, problematical by potentially discounting the lived experiences and knowledges of those who are subject to the analytical gaze. Such knowledge exclusions represent a form of ableist oppression which daily confronts disabled people and is debilitating precisely because it gives them little or no opportunity to contribute to the theorizations of their own subjective status.

This chapter provides a critical overview of such debates and disputes while considering how issues of access, disability, and the built environment ought to be theorized. In particular, in rejecting the pathological underpinnings of the medical model, the chapter seeks to place disablism in the built environment in a broader materialist framework which suggests that a reconstruction of the theorization of disability and the environment requires, as a first step, to place the body in context, to recognize the interplay between the physiological condition and the wider material conditions of existence. In particular, in seeking to move away from essentialist and/or reductionist conceptions of disability and access, I seek to develop the proposition that the experiences of people with disabilities in the built environment should be conceived of as deriving from what Laws (1994a), following Young (1990), terms the multiple and cumulative causes of oppression. In doing so, it is a recognition, following Graham (1990), that the social identities, status and locations of disabled people have a multiplicity of constitutive causes and are, as Laws (1994b) suggests, grounded in prevailing social structures.

In developing such themes, I divide the chapter into four. The first section considers some of the problematical stereotypes and mythologies which have been powerful propagators of what disability is and highlights the role of what Young (1990) terms cultural imperialism, or the dominant group values, in producing and reinforcing systems of systematic discrimination against minority groups. In a second section I develop a critique of the dominant discourses on disability, especially those which have been derived from the medical and/or rehabilitation model of disability. In a third part, I argue that the reductionism inherent in the medical approach to disability requires redress which can be achieved, in part, by placing conceptions of impairment in their socio-cultural contexts. Accordingly, this section of the chapter evaluates approaches which seek to understand impairment as a socio-culturally defined and derived phenomenon. In a final section I consider the possibilities for a non ableist theorization of disability.

STEREOTYPES AND MYTHOLOGIES: SITUATING DISABILITY

Throughout western societies, and beyond, bodily impairments and differences have been a source of social oppression. Racism, in its most obvious form, is premised on a categorization of difference by colour of skin, while, as chapter 1 has documented, disabled people are the poorest in western societies by being discriminated against by employers precisely because of their perceived bodily incapacities. Indeed, having an obvious and/or visible disability is tantamount to challenging societal norms of what bodily health should be, and, as Fine and Asch (1988b) have commented, society has tended to react by seeking to ignore, abuse and hide away those who seemingly do not conform to the prescribed standards of physical attractiveness and functional independence. Indeed, chapter 1, in invoking Young's dimensions of social oppression, conceived of such reactions as part of a wider system of structures defined in and through 'cultural imperialism', or where the dominant meanings of a society render the particular perspectives, or practices, of disabled people invisible and, by implication, irrelevant.

Yet, where societal consciousness of disability is expressed, reactions tend, as Hahn (1988) has noted, to fall into forms of aesthetic and/or existential anxieties. While the former is characterized by people's fears of those with 'unattractive' bodily attributes, the latter is related to perceived, personal threats of the possibilities of 'becoming like that', of acquiring a permanent and debilitating disability. The manifestation of such fears has, historically, taken many forms and the dominant cultural expression of the aesthetic anxiety in western societies has revolved around the notion of the 'body beautiful' or what Hahn (1988) has characterized as the pervasive cultural emphasis on personal attractiveness. As Fisher (1973) has argued:

> despite all the efforts invested by our society in an attempt to rally sympathy for the crippled (sic), they still elicit serious discomfort. It is well documented that the disfigured person makes others feel anxious and he (sic) becomes an object to be warded off. He is viewed as simultaneously inferior and threatening. He becomes associated with the special claims of monster images that haunt each culture (p. 75).

Hahn (1988) extends this in noting how aesthetic anxieties underpin the portrayal of disabled people in pejorative terms and, with regard to disfigured women, the denial to them of ascribed social roles. Moreover, as Blackwell et al (1988, p. 307) have documented, the sense of rolelessness of disfigured women is 'reinforced by social stereotypes that such women are inappropriate as mothers or sexual beings', that somehow their abnormal appearance is a visible manifestation of their inner, obviously distorted and tortured, selves.

Such (bodily) associations, as Hahn (1988) and others note, reflect societal stereotyping which is premised on the notion of the biological inferiority of people with disabilities. In particular, such conceptions are, as Bowe (1978) suggests, part of a wider transcendental historical legacy whereby quite contrasting societies have marginalized those with perceived congenital conditions (Scheer and Groce, 1988). A range of literature, for instance, tends to suggest that in pre-industrial societies people with disabilities were cast out of their communities or even killed (see Scheer and Groce, 1988, for an account of such themes). As Bowe (1978) recounts, survival often depended on physical strength, agility, and sensory acuity, while many societies believed that disability was supernaturally inspired. Moreover, as Fine and Asch (1988a) note, many states in the USA, until recently, had laws which banned disabled people with records of epilepsy and psychiatric disorders from marrying, while there is a plethora of evidence to suggest that parts of the medical profession still try to encourage women with disabilities to be sterilized for fear that they might reproduce their 'disease' in the child to be born.

Societal actions of this type are indicative of forms of oppression which persist, as Young (1990) notes, through unconscious assumptions and stereotypes and group related feelings of nervousness and aversion. As Young argues:

> group oppressions are enacted in this society not primarily in official laws and policies but in informal, often unnoticed and unreflexive speech, bodily reactions to others, conventional practices of everyday interaction and evaluation, aesthetic judgements, and the jokes, images, and stereotypes pervading the media (p. 134).

Contemporary ableism is experienced by many disabled people less as a form of overt discrimination, like the segregation into asylums, but more as what Young refers to as the 'avoidances, aversions, and subtle separations' (p. 142). Those with scar tissue on their face and/or visible disfigurements, for example, become a source of embarrassment for those who gaze at them, and the stunned silences, while people wait for the person with a stammer to finish their sentence, convey the inability of society to accept the multiplicities of the body. Thus, for people with disabilities, societal signals – the body language, averted gazes, the signified discomforts – are all indicative of an abjection or what Young describes as the 'feeling of loathing and disquiet the subject has in encountering certain matter, images, and fantasies – the horrible, to which it can only respond with aversion, with nausea, and distraction' (p. 143).

Yet, far from being a 'natural' phenomenon evoking similar responses in different societies, disabilities were, and still are, conceived of in quite diverse ways (Gleeson, 1995; Hahn, 1988). Such socio-cultural differentiations indicate that disability cannot be collapsed to a fixed measure or state of a person's physiology, somehow a state of naturalness, but rather has its roots in specific socio-spatial and temporal structures. For instance, in pre-modern societies, as Scheer and Groce (1988) have documented, people with disabilities in specific places were often integrated into particular community roles and, as Higgens (1992) has commented, 'rarely were infants who survived past prenatal period killed and the killing of elderly disabled was rare . . . instead, pre-industrial families and communities often felt obliged to take care of their members' (p. 191). In particular, it was common for people with disabilities to play active and important roles in their communities and, as Scheer and Groce (1988) note, 'in the highlands of Mexico blind male elders weave nets, sort produce, tend the gardens, and socialize the young by interpreting their dreams' (p. 29–30).

Moreover, as Hahn (1988) has documented, in small scale rural societies, community and family obligations were underpinned by forms of care and social acceptance of disabled people which only began to rupture with the emergence of the modern industrial period. As Hahn recounts, the onset of industrialization, of the segregation of work from home, and of the emergence of an ethos that only those who were 'fit' to work were of value, were important in transforming the social and economic status of many individuals with disabilities. The production of disability as abnormal, deviant, somehow residing in and of the person, began to reach its apogee during the rise of industrial capitalism and, as Oliver (1990) notes, increasingly the idea of disability as part of a 'natural' state began to take hold. This materiality, defined in and through the evolving capital-labour wage relation, was crucial in 'sorting' people into different types, and those categorized as 'disabled' acquired a label of 'inability to perform'. As a consequence, they were considered to be more or less useless (somehow 'unable') in the emergent systems of productive organization and were, accordingly, segregated into special institutions, hidden and displaced from the mainstream of society (see Oliver, 1990, for an extended discussion of this point).

This, then, suggests a need to interpret, in part, disability as a materialist phenomenon, linked to the changing nature of economic relations in society. Indeed, as Oliver (1990) notes, the emergence of modern discourses about disability was derived from the breakdown of collectivized forms of work organization and a particular language powerfully denoted the emergent categorizations, of those who were worthy or unworthy, the productive and the unproductive, and the able and the disabled. For Foucault (1980), the emergence of industrial capitalism generated a new conception of the body, or the idea that

> the body – the body of individuals and the body of populations – appears as the bearer of new variables, not merely between the scarce and the numerous, the submissive and the restive, rich and poor, healthy and sick, strong and weak, but also between the more or less utilizable, more or less amenable to profitable investment, those with greater or lesser prospects of survival, death and illness, and with more or less capacity for being usefully trained (p. 172).

Indeed, the centring of functional categorizations of this type generated a new dilemma for the state in terms of what to do with the 'incapacitated' or those deemed to be 'unable' to work. It posed a problem of care and control while underpinning stereotypical conceptions of the nature of the incapacities, that somehow disabled people were, as Wolfensberger (1975) notes, 'childlike', in need of control, management and spiritual guidance. Indeed, the material manifestations of such attitudes, as the next chapter documents, were institutionalized in a range of state policies and the 1913 Mental Deficiency Act in the UK, for instance, was one of the first major pieces of state intervention for people with learning difficulties. Yet it wholly reinforced the negative images of such individuals by its statutory declaration that they were 'mentally deficient', 'idiots', 'imbeciles' or 'feeble minded'.

Yet, while the materiality of disability, its production and reproduction in the context of economic relations, has been, and continues to be, an important locus through which we have to understand, in part, the status of disabled people in society, it is clear that the ableist values and attitudes of western societies are deeply embedded in a range of socio-cultural assumptions about impairment, disfigurement and disability. Thus, in a pointed critique, Fine and Asch (1988b) unpick a range of stereotypical assumptions about disabled people. One of the most enduring is the assumption that a disability is wholly derived from a person's physiology, that the 'disability and person are seen as synonymous' (p. 2). Indeed, aspects of medical sociology and social psychology retain the idea that disability is reducible to the purely physical and/or mental, the notion that a disabling state is an independent (determinate) variable. As Fine and Asch note, the possibilities that disability might be influenced by a person's race, gender and/or class status are wholly ignored in a portrayal of disability as 'the variable that predicts the outcomes of social interactions' (p. 9).

Such conceptions have underpinned what little theorizations there are of disability and access in the built environment, where it is generally assumed that physical impairment is the cause, and consequence, of a person's inability to interact with their social and physical environments. Indeed, throughout the last fifty years, social psychological and medical research, which has dominated the study of disability, has reaffirmed the notion that somehow social structures are a given, a fixed entity, while, as Fine and Asch (1988b) observe, encouraging disabled people to accept counselling and appropriate (medical) treatment to enable them to cope with, and accept, a world which will inevitably frustrate them. Thus, Barker's (1977) research on disabling environments concludes that it was more or less impossible to legislate for the removal of all restrictions on the physically deviant in a world constructed for the physically normal. The ultimate adjustment must 'involve changes in the values of the physically disabled person' (p. 37, quoted in Fine and Asch, 1988b, p. 9). This is echoed by others, most notably by Kasprzyk (1983) and Golledge (1991, 1993), who assert that the obstacles facing disabled people are, first and foremost, their impairments rather than the physical barriers of the built environment.

In part, such attitudes are linked to notions of aversion and abjection mentioned earlier (see Kristeva, 1982). In particular, dominant group cultures, historically, have tended to react against any grouping perceived as a threat to (ableist) conceptions of identity and, as Young (1990) notes, it reflects a deep rooted xenophobia, 'structured by a medicalized reason that defines some bodies as degenerate' (p. 72). Ableism has developed, in part, from such notions and has been sustained because such despised and marginalized peoples have been objectified or, as Young (1990) argues, 'scientific, medical, moral, and legal discourse construct these groups as objects, having their own specific nature and attributes, different from and ever against the naming subject, who controls, manipulates, and dominates them' (p. 146). One of the classic illustrations of this is Joseph Merrick, more commonly known as 'the elephant man', a person with congenital disfigurements who was incarcerated in a Victorian freak show before being 'rescued' by the medical profession. Yet, as accounts have shown, the medical profession utilized Merrick as an object of curiosity, a deviation, to be displayed for the (medical) curiosity of 'scientific' endeavour (Darke, 1993). In this sense, Merrick was being (re) presented as part of 'the public gaze' and, in Darke's terms, was part of a contemporary ethos dehumanizing and objectifying particular subjects 'rather than representing abnormality/disability as human and valid in itself' (p. 340).

Such forms of cultural representation illustrate the ways in which particular human beings, especially in late Victorian society, were treated as something other than human, 'as potential exhibits in what was a cross between a zoo and a museum' (Darke, 1993, p. 339). Indeed, such depictions suggest that ablebodiedness was being extolled as an absolute human type while, as Shakespeare (1993) notes, disability was being treated as an 'otherness'. To quote Jordanova (1989):

> the treatment of the other as more like an object, something to be managed and possessed, and as dangerous, wild, threatening. At the same time, the other becomes an entity whose very separateness inspires curiosity, invites enquiring knowledge. The other is to be veiled and unveiled (p. 109).

The notion of 'otherness', then, is a useful formulation and has had its most explicit formulation in a range of continental philosophy from Lacan (1977) to Barthes (1975) through, more recently, to Foucault (1980). As Shakespeare (1993) notes, De Beauvoir (1943, 1976), in classic Hegelian terms, situated the notion of otherness by declaring that each consciousness seeks the death of the other. In this context, De Beauvoir was referring to women as 'the second sex', as generalized beings rather than as individuals, or that masculinity is the absolute form and type and women, like people with disabilities, are 'out there', estranged, strange and marginalized.

There are also senses in which disabled people are presented as victims left to cope with their suffering which is infused with self blame. As Fine and Asch (1988a, 1988b) note, it is assumed, here, that the disability, in itself, constitutes the victimizing experience. Indeed, the associations of decay and death with disability, are evident in the literature and, as Dickenson (1977) has commented, 'patients must be allowed to come to terms, they must grieve and mourn for their lost limbs, lost abilities, or lost looks, and be helped to adjust to their lost bodily images' (p. 12). The metaphor of death, underpinning Dickenson's observations, seems to signal a tragic finality, as though life beyond the 'normal' body is one which is barely worth living. As Gooding (1994) notes, this form of reaction is an integral element of socio-cultural definitions of disability in western societies based 'on a combination of patronization, pity and discomfort . . . although fear, hatred, and hostility are also prevalent' (p. 2). It also draws attention to the impairment, so projecting the idea that it is this, in itself, that provides disabled people with their core self understanding.

In particular, a common aspect of popular culture is the reduction of a disabled person's self concept to their impairment, and disabled identities, as portrayed in media, commonly evoke images of 'more than or less than human, rarely as ordinary people doing ordinary things' (Oliver, 1990, p. 61). Images range from the portrayal of disabled people as sinister or evil to that of hero, as a super-being struggling against the odds (Johnson, 1994; Nelson, 1994). Likewise, literature reinforces pejorative images of people with disabilities with, as Nelson (1994) recounts, fairy tales showing disabled people as afflicted and evil, while William Shakespeare's play, *Richard III*, depicts bodily deformity as a cause and consequence of an evil pathology. In so doing, such conceptions tend to reinforce the 'personal tragedy' view of disability and, even where depictions of disability are positive, such as someone struggling against blindness or paralysis, the received stereotype is, as Nelson (1994) notes, one of the disability being the major focus of the individual's life. In this sense, the cultural and material appropriation of disabled people's lives and experiences, of the types documented above, both reflect and reproduce ideologies of individualism and, in doing so, play a significant role in reinforcing the cultural imperialism of ableism.

DISABLISM AND 'THE DESERVING POOR': A SOCIAL PATHOLOGY OF DISABILITY

As the previous section has argued, the dominant ideology is to view disability as personal with the real difficulties faced by disabled people as the result of having an 'impairment'. Indeed, despite a decade of protest and campaigning by disabled groups for the recognition of their rights to participate fully in society, societal attitudes are still clearly framed by conceptions of the 'deserving poor', whereby a state of disability is understood as a product of a medical condition and/or an individual pathology (Barnes, 1990; de Jong, 1983). In particular, many policy makers have accepted the prevalent assumption that programmes for the removal of physical barriers are necessitated by the functional limitations of people with disabilities. Yet such matters tend to be regarded as technical issues or as compensatory measures merely to assist a small and economically insignificant population with special needs (Hahn, 1986b, Speare *et al*, 1991). As Hahn (1986a, 1986b) suggests, such suppositions continue to perpetuate a belief that providing accessibility for disabled people is simply the extension of privilege or even charity.

Such ideas are espoused by the functional limitations paradigm which asserts that the most significant difficulty with disability is the loss of physical or occupational capability. This perspective argues that the disability resides exclusively within the individual, that it is reducible to the nature of the impairment and is treatable much as a doctor would attempt to cure a patient's disease. A range of research clearly highlights the medicalization of disability as a significant factor in society's marginal treatment of disability issues (Martin and Elliot, 1992). Oliver (1990) notes that the World Health Organization (WHO, see Wood, 1981) still cling to a medical classification of disability which sees a disabled state as a form of disease and/or abnormality. Indeed, its definitional basis, as Figure 2.1 depicts, tends to take the concept of 'normality' for granted in defining disability as 'not being able to perform an activity considered normal for a human being'. Yet, as Oliver and others have argued, there is little consensus on what constitutes a state of normality, while the WHO's definitions of disability fail to recognize the situational and cultural relativity of how normality is understood.

Indeed, as Soder (1984) notes, as long as the environment consists of social roles considered to be normal, the inability of any individual to conform puts them in a disadvantaged position and thus creates a handicap. In this way, the medical approach is conserved, since changes must be brought to bear on the

Figure 2.1   The World Health Organization's international classification of impairment, disability, and handicap

(The WHO utilize a threefold categorization by which to define disability and its definitions have become the standard bearer which most countries in the world have adopted)

Category	Definition
Impairments	'disturbances in body structures or processes which are present at birth or result from later injury or disease'
Disabilities	'limitations in expected functional activity or as restrictions in activity due to an underlying impairment'
Handicap	'difficulties in performing activities of daily living, like walking . . .'

Source: Adapted from Woods, 1981.

individual rather than the environment. Part of the problem with this (ableist) conception is the way in which it treats disability as uniform and homogeneous, reinforcing the notion that there are two discrete types, the able-bodied and the disabled, with the former leading a much more enriched existence than the latter. As Figure 2.2 indicates, the medical model projects a dualism which categorizes the able-bodied as somehow 'better' and 'superior', literally 'more able'. In contrast, the disabled are conceived of as 'unable' and requiring help or the application of professional medical services. Such conceptions, in total, conceive of the body as a biological entity, something which is natural and pre-given, a fixed state of being.

Figure 2.2 Social stereotyping and conceiving of the 'ablebodied' and 'disabled' as unequal and opposites.

Ablebodied	Disabled
Normal	Abnormal
Good	Bad
Clean	Unclean
Fit	Unfit
Able	Unable
Independent	Dependent

Yet, as I have already intimated, people with disabilities do not readily conform to 'type' and, as Scotch (1988) has indicated, 'disability', as a unifying concept 'that includes people with a wide range of physical and mental impairments, is by no means an obvious category' (p. 159). In Sayer's (1984) terms, it represents the classic case of a chaotic conception, or a notion which is unable to describe the complex, contingent nature of physical and/or mental impairments, while possessing little or no analytical value helping to explain the processes which give rise to the socio-cultural (re) production of the very state(s) described as 'disability'. Indeed, the WHO approach is taxonomic as though disability represents a series of 'types', of physical and/or mental end states, rather than comprising part of a dynamic socio-cultural, and physiological process (see Figure 2.1). That is, there is little acknowledgement of the social construction of disability as something which reflects specific societal values, fears, prejudices, even ignorance, or, as Oliver (1990) says, 'it conserves the notion of impairment as abnormality in function, disability as not being able to perform an activity considered normal for a human being, and handicap as the inability to perform a normal social role' (p. 4).

In particular, part of the problem with this perspective is the way in which it fails to understand how disabled people are socialized into particular ways of being, of accepting their 'inferiority' by society, and behaving in ways which seemingly conform to the expectations and stereotypes which have been handed down. As Abberley (1993) notes, the range of disciplines, from medical sociology to social psychology, still retains the notion that disabled people are abnormal in the sense that their impairment can only be explained in terms of a deviation from a 'standard norm', that they are the problem for deviating from it! Yet, as Oliver (1990) and Abberley (1993) have noted, if the notion of abnormality is placed in a different type of context, 'not in disabled people but in the society which fails to meet their needs', then a different type of understanding of 'normality' is generated (Abberley, 1993, p. 111). As Abberley convincingly states, 'our abnormality consists in us having . . . a particular and large set of our human needs unprovided for, or met in inappropriate and disempowering ways . . . it is in this sense, of having an abnormal number of our normal needs unmet, that I think it is right to speak of disabled people as not being normal' (p. 111).

Yet the notion of disability as an individual abnormality retains a powerful hold over social theory and has perpetuated a range of research which wholly abstracts from conceiving of the body in its socio-cultural contexts. Thus, Anderson and Clarke (1982) show how low self esteem is a characteristic of adolescents, while Kasprzyk (1983) indicates how despondency is a more or less recurrent state among people with spinal injuries. Moreover, experimental psychologists, in attempting to simulate disabilities, have concluded that people with disabilities arouse anxiety and discomfort in others and, as a result, are socially stigmatized (see Kasprzyk, 1983). Yet, as Fine and Asch (1988b) comment, such experiments tell little or nothing about how disabled people engage in meaningful social interactions and there is an overarching reducibility in the conception, in that 'disability is portrayed as the variable that predicts the outcome of social interaction when, in fact, the social context shapes the meaning of the disability in a person's life' (p. 19).

Indeed, as Fine and Asch (1988b) note, such conceptions sustain the idea that people with disabilities are somehow weak and dependent, that their 'biological condition rather than the environment and social context makes one-way assistance inevitable' (p. 6). As Fine and Asch conclude, such assumptions perpetuate a negative and demeaning image of disabled people, that somehow their physical incapacities, in themselves, are debilitating to the extent they are incorporated into most other spheres of their social and economic lives. In summarizing such perspectives, Fine and Asch lucidly note that:

It is the disability, not the institutional, physical, and attitudinal environment that is blamed for role changes that might occur. The person with a disability may (initially, or always) need physical caretaking, such as help in dressing, household chores, or reading. It must be asked, however, whether such assistance would be necessary if environments were adapted to the needs of people with disabilities – if, for example, more homes were built to accommodate those with wheelchairs . . . if technological aids could be developed to convert the printed word into speech or braille and were affordable to all who needed it . . . the physical environment as an obstruction remains an unchallenged given (p. 14).

And because the physical or built environment remains a given in the medical theorization of disability, there has been a tendency for disabled people to be stigmatized or, as Barnes (1991) argues, the association of disability with stigma wholly reinforces the notion of disability as an individually derived problem. As Oliver (1990) notes, the origins of the idea of disability as stigma relate to Goffman's conception that stigma is a form of societal branding where individuals transgress the norms or values of society. Indeed, while stigmatized identities are derived through interpersonal interactions, for Oliver (1990) the explanatory utility of the idea is problematical because

while stigma may have existed in all societies, in ancient ones it was inflicted because of some transgression or other; in modern societies, the stigma itself was the transgression. In both kinds of societies, stigma implied moral opprobrium or blameworthiness (p. 65).

Underpinning this is what Dalley (1992) terms the normalization thesis or that perspective which notes that because disabled people are labelled they are devalued and that the mechanism to reverse this is one of 'normalization' or the removal of pejorative labels and/or social categories. For Wolfensberger (1983) normalization must be the creation, support, and defence of valued social roles for people who are at risk of social devaluation yet, as Dalley (1992) notes, this solely concentrates on the 'roles' that people occupy rather than on people as 'the persons that they are'. As Dalley suggests, the essence of normalization is that of social conformity, the idea of the reducibility of diversity and difference to a specific type or, as Carver and Rhodda (1978) have characterized it, 'the focus is firmly on the rehabilitatee with the objective of re-modelling him (sic) as closely as possible to the functional semblance of an average person' (p. 10).

However, as Dalley (1992), and others, have noted, there is a range of problems with the normalization thesis, not the least of which is the assumption that incorporation into mainstream society is a good thing. As Oliver (1990) has argued, underpinning such perspectives is the idea that all people should be returned to the state of the fit, able-bodied, individual or, as Finkelstein (1988, p. 4) has commented, 'the aim of returning the individual to normality is the critical foundation stone upon which the whole rehabilitation machine is constructed'. Moreover, as Szivas (1992, p. 112) argues, it is also assumed by the mainstream that 'to be attributed value disadvantaged groups should aspire to fulfil society's idealized norms'. Yet, as Dalley (1992) notes, this then generates a context whereby the advantaged, dominant, groupings define what is or is not to be valued. In this sense, I concur with Szivas (1992), who notes that the notion of normalization conceives of difference negatively 'making it impossible to avoid disaffiliation . . . and shame' (p. 113).

Thus, while the emergence, post second world war, of rehabilitation services and programmes, especially in the USA, held up some promise of overturning medical conceptions of disability, they tended to reinforce the notion of the disabled person as an inferior being. In particular, the history of rehabilitation programmes, both in the USA and the UK, has been driven by professional elites shaping and constructing the meanings of disability around technical, socio-psychotic, and medical concerns, which, as Trent (1994) has argued, deflected attention from questions of power, status and, ultimately, control. As Slee (1993) notes, the underlying policies tended to generate an objective opacity, a form of neutrality, which reduced issues of integration and 'normalization' to policies of technical adaptation. Thus, as Slee has indicated, the integrationist policies of education authorities in the UK are similarly disablist in reducing the debate over integration (or its absence) to questions concerning disputes over resources. That is, given the resources, integration will occur. Yet, as Slee comments:

> such debates sustain the flawed notion that integration is simply a technical issue to be achieved via the deployment of special equipment and personnel to regular schools . . . it deproblematizes integration through the absence of appreciation of the social construction of disability . . . (p. 359).

In seeking to move beyond such limiting, and limited, perspectives a range of alternative conceptions has been mooted. One of the more significant is Emener's (1987) empowerment model which is concerned with addressing how the professional system of rehabilitation might empower disabled people to gain control over their lives. Its real strength is the departure from a model of functional impairment to the notion that a disabled person should have equal opportunity to maximize his or her potential and is deserving of societal help in attempting to do so. Yet, as an approach to theorizing about the nature of disability, Emener's model is weak in several respects. Foremost, it is wholly based on supporting the role of professional and technical elites in delivering and underpinning support for disabled people and, consequently, fails to theorize how institutions are implicated in the propagation and perpetuation of disablism. Emener's thinking is revealing here because he adopts a paternalistic notion that, while people with disabilities must be empowered, the locus of control, as Emener argues, is critically contingent upon rehabilitation (p. 1). This, then, reinforces the idea of the efficacy and importance of professional control.

Indeed, the idea of empowerment, underpinning the approach, is limited by its failure to propose the means of combating adverse institutional attitudes and responses towards impairment. Moreover, the perspective conceptualizes people with disabilities as consumers, or rehabilitation clients, being acted upon and lacking the capacity to transform their lives without the help of the professional bodies, and, as Hahn (1988) notes, Emener's notion of empowerment does little to transform the iniquitous, and hierarchical, relations between professionals and disabled people, leaving the 'major principles and content of . . . rehabilitation counselling relatively undisturbed' (p. 41). In particular, the perspective maintains the principles of an ableist cultural hegemony while reinforcing the idea that somehow the disabled person is still to blame. In this sense, the notion of empowerment here is duplicitous, a misnomer, while its underlying social theory fails to recognize how rehabilitation, in itself, serves to reproduce many of the social relations of ableism.

BEYOND SOCIAL PATHOLOGY: CONTEXTUALIZING DISABLISM

The contrasting approaches to the theorization of disability indicate a gradual move from the representation of disability as an individual pathology towards a social constructivist model, or one which situates our understanding of disability in a wider context of social and political relations. In particular, the failures of the medical and/or rehabilitation model of disability have led a range of authors to situate our understanding of access issues in the built environment in the values, attitudes, and policy programmes of institutions and their actors. Indeed, as Hahn (1986a) suggests, one of the keys to understanding disability in the built environment, and elsewhere, rests with an exploration of the determinate institutions and 'the solution must be found in laws and policies to change the milieu rather than in unrelenting efforts to improve the capacities of a disabled individual' (p. 276).

This approach towards setting disability in context is exemplified in what Hahn (1988) has termed the minority group model of disability, a perspective which embodies both social constructivist and creationist views of disability. At the broadest level, the minority group view situates disability in the wider structural, external environment, denying that it is explicable as a consequence of some personal defect or deficiency. Hahn notes that the 'minority group' model offers a means of transcending the limitations of medical models of disability by focusing on socio-political distinctions that see people with disabilities as the 'product of interaction between the individual and the environment' (p. 40). As Hahn argues, this perspective does not regard disability as a personal deficiency but as the result of the social conditioning of disabling environments. In this sense, the analytical focus on disabled people is switched from a concern with the internal, individual, defects of the person (a socio-pathological approach) towards the wider structural, or external, underpinnings of a disablist society, of its values, attitudes and public policies.

The implication of the minority approach, or what Hahn (1988) terms the socio-political definition of disability, is that it develops a social constructivist position by situating disablism within the oppressive and coercive ableist attitudes of society, attitudes reinforced and perpetuated by the practices and discourses of the dominant institutions. As Hahn suggests, the underlying political message is of the 'need to transform formerly devalued attributes into positive sources of dignity and pride and entailing self management skills and a positive self concept' (p. 40). This, then, recognizes that, first and foremost, 'attitudinal discrimination is the major problem facing those with disabilities' (Hahn, 1986a, p. 276). In this sense, the real barriers to access, from the social constructivist position, are not the physical barriers in themselves, but the wider attitudinal strictures of prejudice and discrimination against people with disabilities.

Yet the approach is problematical in a number of ways because, by locating sources of oppression solely in 'attitudes', there is little sense of their social location or origins, or of how attitudes are translated, if at all, into oppressive actions. While not denying the interplay between ideologies, values, and actions, the real weakness of the social constructivist part of the minority model is the absence of any account of the socio-political contexts within which values and attitudes arise, and of their transformative capacities. Indeed, the interplay between attitudes, values, and material practices is difficult to specify and has the capacity to reduce the discriminatory practices of ableism to a 'state of mind', or what Gleeson (1995) refers to as 'a discriminatory set of beliefs which are imposed upon different, if essentially, normal people' (p. 20). Thus, such notions are idealist because they fail to situate ideas and values socially, culturally, or historically, and are problematical for the very reason that they, as Hevey (1991, p. 14, quoted in Gleeson, 1995, p. 20) claims, take the material world as a given.

The other interrelated element of the minority group approach, the social creationist perspective, locates the source of social oppression of disabled people in the socio-institutional practices of the dominant professional groups. As Hahn (1986b) argues, public policy is a reflection of pervasive attitudes and policies, and disablist attitudes, in themselves, have transformative capacities in influencing the policies and practices of institutions. In this sense, disablism is seen as something which is locked into, and located within, the behaviour of powerful organizations and institutions. As Young (1990) notes, welfare states, post 1945, have been and still are preoccupied with notions of normalization, of a de-differentiation process which is wholly subversive of (disabled) 'identity as difference' (see chapter 3 for an extended discussion of these debates). In this sense, the social creationist conception represents an advance on social constructivism because it places values and attitudes in a material context of socio-political practices while recognizing that institutional domination, or the prevention of people from participating in determining their actions or the conditions of their actions, is a key structural facet of administrative and welfare control over the lives of disabled people.

Yet the perspective is also problematical in reducing the propagation of ableist attitudes solely to the realm of socio-institutional practices, or to public policy. That is, it tries to say that disability is what is defined by public laws and programmes, what Birkenbach (1993) refers to as socially constructed reality rather than a biological fact. Moreover, the approach propagates the idea that discrimination against people with disabilities should be eradicated and equal opportunity policies instigated as one of the measures towards their emancipation. Yet, as Birkenbach argues, surely the socio-political model must recognize that there is a physical state, a physiological status which really negates any possibility of people with disabilities being afforded equal opportunities and treatment in that their very (physical) differences demand a difference in the way society responds to them and their social, human, and physical needs. Indeed, the physicality of the body is too often ignored in such perspectives (see next section).

In particular, elements of disability protest in the UK have formulated political strategy around the conception that the interrelatedness between values, attitudes and socio-institutional practices which exclude people with disabilities from the built environment have to be challenged in ways which transcend legal tinkering or efforts to 'socially engineer' a 'solution'. Thus, the development of 'crip-politics, for example, in the USA, and of political movements which seek to emphasize impairment as difference are, in essence, saying that we want to be recognized for what we are rather than what you (through your legislation) want us to be.[1] Indeed, Lane (1995), discussing the construction of deafness, supports elements of social creationism for its critique of the service professionals, especially of the way in which they service 'not only their clientele but also themselves and are actively involved in perpetuating and expanding their activities' (for their own legitimacy) (p. 174). Others, like Morris (1993), want to 'bring the body back in', to demonstrate that the physical and/or mental impairment determines an individual's behaviour (independent of wider social constructs).

In seeking to move beyond the idealist and also institutionally grounded conception of disability, an important development in the theorization of disability, albeit with significant weaknesses, is the location of disability within a materialist perspective. Its real strength is its positioning of disability historically, noting how states of disability are (re) produced and/or made and not the consequences of an impairment. Indeed, the distinction between 'impairment' and 'disability' is crucial in that materialists conceive of the former as the absence in total, or in part, of some physiological function, while the latter is the 'socially imposed state of exclusion or constraint which physically impaired people may be forced to endure' (Gleeson, 1995, p. 12). In this sense, as Gleeson notes, impairment is indicative of a particular bodily state, a specific physiology which is only ascribed a particular social meaning in particular socio-cultural contexts. As Gleeson comments, 'impairment can only be understood – historically and culturally – through its socialization as disability or some other (less repressive) social identity' (p. 12). Yet, this conception seems reductionist because impairment is also a bodily state and, in its physical state, has the capacity to create (physical) pain and discomfort which need not necessarily be socially and/or culturally reproduced or ascribed.

Thus, Oliver (1990), Gleeson (1995) and Stone (1984), to name but a few, tend to identify the 'socialization of disability' as reducible to the material conditions of society. At its base is the idea that the social oppression of disabled people is related to the value of their labour power, or capacity to work, that their inability to perform 'normal' work more or less excludes them from the labour market and, consequentially, from a regular wage. For Stone (1984), for instance, the category 'disabled' emerged in the nineteenth century as the state's response towards sorting out the ablebodied from the disabled, or those with the (measurable) capacity to work from those without, yet this (materialist) conception of disability seems functionalist. It also conceives of class as the key social variable as though the socially oppressive nature of disability is best viewed through the lens of conflicting class relations. Likewise, Gleeson's framework is essentialist by arguing that the primary determinant of disabled people's oppression is their economic status and exclusion from the labour market, that class struggle, in and of itself, is the essential determinant of societal transformation.

Oliver (1990) takes a similar stance in conceiving of people with disabilities as little more than the consequences of material relations yet, as Gleeson (1995) and Tomlinson and Colquhoun (1995) note, in rejecting psychologically inclined explanations, Oliver ignores the determinant powers of culture, representations, and their associated meanings. While, in part, this is slightly unfair to Oliver who does recognize socio-cultural dimensions in the construction of disability, there is an impression from his work that the problems associated with disability will disappear if the underlying (material) relations of a disablist society were to be transformed. Thus, the logic seems to say that if one adapts the social and physical world for disabled people, then the disabilities will dissolve, yet this still leaves the thorny issues of the body (and the impairment) and of the possibility, that by objectifying bodily experiences in a social model, that the subjective, real, experiences of, for example, physical incapacity and pain will be ignored or just dismissed. Indeed, even if the oppressive social relations of disablism were to be transformed would that necessarily remove the physicality (the reality) of the body?

Also, as Laws (1994a) notes, if one acknowledges that class is not the essential cause of oppression, that oppression can take many different forms, then a different type of conceptual schema is required. In particular, Laws and others suggest that the diversity, differences and heterogeneities of socio-political structures and processes should be acknowledged, that the lived materialities of people have to be characterized by their varied and complex struggles and contestations relating to social reproduction. As Laws (1994a) notes, the real challenge is to regard oppression as a social relation, not as an ossified essentialist category reducible to a singular dimensional form. In this sense, the socio-spatial nature of disability is something which is not reducible to the technical or the physical, but is located in wider structural systems of social and economic oppression. For instance, while the built environment plays an active role in shaping the experiences of people with disabilities, the environment, in itself, does not cause the oppression which disabled people face but is constitutive of the structures which reflect and feed from the wider socio-political structures of ableism. Indeed, as Gleeson (1995) powerfully argues:

a great entrapment for policy practice is the tendency to reduce the dynamic socio-spatial nature of disablement to a built environment problem, so that disability merely becomes a problem of physical inaccessibility in a thoughtlessly designed built environment (p. 20).

This, then, focuses attention not so much on how the configuration of buildings causes physical and/or social dislocation, but on how and why such configurations were produced in the first instance. As Gleeson (1995) notes, marginal transformations of environments can do little more than ameliorate the underlying effects, that to focus on how planning regulations and/or statutes can reduce the mobility of disabled people is to reduce the solution to a technical and/or spatial fix while ignoring the social and political processes which created them. Yet, while this is a useful observation, it stops short of theorizing the processes responsible for the multiple nature of social oppression underpinning the lives of people with disabilities and, in particular, ignores the centrality of the body in the socio-cultural (material) constructions of disability.

THE POSSIBILITIES FOR A NON ABLEIST THEORIZATION OF DISABILITY AND ACCESS

The intellectual limitations of much social theory about disability has led a range of social theorists to call for the development of a non ableist, non essentialist, sociology based upon setting sensory feelings and physiological impairments in their socio-cultural contexts (Birkenbach, 1993; Butler, 1995; Dear, 1995; Gleeson, 1995). As Butler (1995) argues, physiological impairments, in themselves, are a constraint on specific types of action and it is impossible to derive a social theory which is dismissive of, and independent from, the situatedness of the body, of its psychological and/or physiological state. Likewise, Birkenbach (1993) makes the powerful case that a non ableist theorization must recognize the 'interactional' character of disablement. By this, he argues that it is imperative that one locates disablement 'in a relationship between a medical and functional problem and the social responses to it' as the only way to escape charges of essentialism and/or analytical reductionism (p. 178). Others, like Pile and Thrift (1995), refer to the body as a 'point of capture', or where personal experiences accumulate and shape the 'being' of the embodied person, while for Hall (1995) the body is 'an active and reactive entity which is not just part of us but is who we are' (p. 10). Thus for Hall and others the body is 'corporeal', 'neither determined by biological or social processes, but absorbing and reacting to social and biological processes' (Hall, 1995, p. 13).

In an important contribution to the debate, Hall (1995), echoing Birkenbach (1993), suggests that the basis for a non ableist theorization of disability can only occur if the dualisms of able/disabled, ability/disability, and normal/abnormal, are dissolved, that is for the fluidity of the concepts to be recognized and for the body to be situated and interpreted as a socio-cultural and biological construction, neither fixed nor unchanging (thus interconnecting social and medical conceptions of disability).[2] Indeed, these dualisms are powerful and fixed conceptions of how to conceive of the body and, as I have already argued, they tend to be underpinned by hegemonic, culturally imperialist, conceptions of embodiment. Thus, the able-body is somehow 'more able', better than, the disabled body, the 'unable', yet this fails to recognize the daily changing states of our bodies, both physiologically and in how we feel about them, to the extent that maintaining a division between two ossified, static, categories, is more or less meaningless. The corporeality of the body, however, indicates that how we feel about our bodies, indeed, how we physically experience them, is temporally/spatially specific, and that there is rarely a constant in the ways in which we receive our bodies and how, in turn, they are received. So, while a person with inflamed facial scar tissue may feel unable (disabled) to 'face' the world, to go out on the streets for fear of the stares, at another moment, when the inflamed tissue has subsided, the interaction between their physicality and 'those outside' becomes transformed, and they are 'abled'.

Thus, at different moments the same person, the same body, is 'abled' or 'disabled' in that the socio-cultural attributions of society towards facial disfigurement will interplay with the individual's 'looks' in producing different gazes, different reactions. Of course, the problem contained within such illustrations is that they maintain the power of the abled/disabled dualism, that somehow people behave in ways which reinforce the wider societal conception of disability as abnormality. However, a notion of fluidity between the two, abled/disabled, is present and, as Hall (1995) suggests, by centring the body in social analysis it then becomes a component in the construction, indeed, our understanding, of the socio-cultural experiences of (disabled) people. However, there is some resistance to this, and the social model of disability is implicated in ignoring the role of the body in socially constructing disability. Thus, in their zeal to assert that disabled people are discriminated against in the labour market, organizations like the Trade Union Congress (TUC, 1993), in echoing a social model of disability, have made the claim that disability is 'caused' solely by societal discrimination. Yet, in claiming this, the TUC divide the physiology, the impairment, from disability, and force a (conceptual) disjuncture which is less than helpful to the person with a disability. Again, the missing ingredient is a conception of the body as a (real) changing biological and/or physiological part of the person that might make a difference to their 'capacity to work'.

Hall (1995) also cites the example of the national charity RADAR (1993) which has claimed that the body is no restraint to employment or, as Hall notes, 'the body is central to these representations of disability, as great efforts are made to deny its role, it is hidden, not allowed into the debate' (p. 23). However, as Morris (1993) has argued, the body can enable or restrain, the pain of a disease is a physical experience with the capacity to debilitate and to reduce a person to a state of complete inability and dependence on carers. Similarly, French (1993) rejects the idea that her visual impairment generates disabilities which are wholly socially created. As she comments, the impairment disables her from recognizing people and makes her 'unable to read non verbal cues or emit them correctly' (p. 17, quoted in Hall, 1995, p. 8). Others, in part, concur with this, and Hall cites the Employers Forum on Disability, a national organization campaigning for disabled people's rights in the labour market, which emphasizes bodily relationships between different employees who, when working beside each other, perceive the 'other' to be very different. As Hall says, their documentation tends to indicate 'that bodies are the same but different, and that difference in body shape/image is important, but also that such aspects of the body are not constant – difference is a changing phenomenon' (p. 22).

Yet there is also a pressing need to develop conceptions of what Dear (1995) has termed 'the body in context', to consider the structural and contingent conditions of its production and reproduction, to interrelate physiological and socio-cultural variables as part of a dialectic. In part, this echoes Bordo's (1995) critique against conceiving of the body as a purely biological or natural form while guarding against a purely non-physiological, culturalist, perspective of the body in context. Indeed, a reconstruction of the body in context is dependent on two conceptual preliminaries being addressed. The first is to sustain the critique against the traditional, ahistoric, foundationalist conceptions of the body as somehow solely being 'derived from nature'. The reductionism inherent in such a formulation is problematical because it sees the body as a neutral, generic, core or as Bordo (1995) notes, the body is conceived of as a sameness 'as though one model were equally and accurately descriptive of all human bodily experience', irrespective of sex, race, age or any other socio-cultural attribute. Yet, as Marx (1969) has noted, the body is much more than a biological phenomenon, it is socio-culturally situated too, and such situatedness is implicated in the explication of bodies in context. This, then, leads to a second, and linked, conceptual problem, of how to transcend what Bordo refers to as the 'old metaphor of the Body Politic presented . . . as a generic (that is, ostensibly human but covertly male) form' (p. 34). This is the problem of the 'otherness' discussed earlier yet, as Bordo (1995) notes, the tendency to explicate 'bodies in context' within a masculinist, disablist, framework is overwhelmingly powerful.

Moreover, as Dear (1995) has suggested, where disability meets social theory is where the body needs to be conceived of as existing in a context of marginality, where marginality is understood as a 'state of being'. For Dear, this conception is defined in and through the interactions of the bodily or physiological condition and the material conditions of existence. Such materiality, for Dear, relates to the interplay between various scales of socio-spatial change including global-local restructuring and the intensification of the power of the global over the local and the subsequent weakening of local employment structures and forms of labour organization and representation. For Dear, and others, this represents one material base where disabled people, already marginalized, have experienced a worsening of their economic status. In addition, the emergence of new regimes of social and welfare control, coupled with the gradual shift towards privatized forms of service delivery, is a significant basis around which the terms of disability and marginalization are being transformed. However, the trick for theorists is to somehow bring the material together with the physical, the physiological with the social.

CONCLUSIONS

The different theorizations of disability are, in their own ways, reductionist and unable to do justice to the multidimensionality of disablement. Whether one takes a biomedical perspective or a materialist position, each, as Birkenbach (1993) notes, 'tries to extend and stretch an important intuition about the nature of disablement in order to distil a complex notion down to one of its core components' (p. 178). Thus, whereas the biomedical approach emphasizes the importance of the physical impairment and the need for a medical response, the social constructivist position situates impairment in a wider context of socio-political relations which discriminate and thereby 'handicap' people with disabilities. In this context, policy ought to be about transcending oppressive socio-cultural values and related institutional practices. Yet, as Birkenbach powerfully notes, such conceptions are inherently weak because they

> deny the interactional character of disablement. Perhaps this is understandable given the limitations of the metaphor: it strains one's imaging powers to try to locate disablement in a relationship between a medical and functional problem and the social responses to it, as the concept of disability requires (p. 178).

Yet the idea that disability is somehow akin to a medical condition is still a powerful underpinning of official attitudes and responses to disability. In the context of the built environment, it tends to reinforce the notion that the body must be 'fixed' to fit the environment, thus emphasizing cure and rehabilitation. Socio-cultural prejudices are ignored, disablism does not exist. The converse is reflected in academic subjects like geography and planning which are largely underpinned by forms of environmental determinism, or the notion that professionals, like architects and planners, can re-design spaces and places to reduce the problems of access and mobility facing disabled people. Thus, as Golledge (1993, quoted in Gleeson, 1995, p. 19) argues:

if society as a whole wishes to provide some semblance of normal independent life for these populations, significant investments must be made both in terms of modifying the environment and in terms of getting information to disabled people (Golledge, p. 70).

This perspective tends to suggest that if you change the physical configuration of the built environment you can change the experiences of people with disabilities. However, such transformations, in and of themselves, will do little or nothing to eradicate the underlying, disablist, values of society or of the institutional structures within which most disabled people have to lead their lives. The reverse is more likely because such perspectives de-politicize the very essence of 'being disabled' as either an individual condition or one connected to the policy and practices of institutions. Wider structural conditions are lost sight of while the body is conceived of (if at all) as ephemeral. Indeed, as Hahn (1988) concludes, while rehabilitation, planning, architectural and other social service programmes have some role to play in creating the conditions for disabled people to achieve some measure of equal opportunities in society, of much greater significance is the pursuit of civil rights and of the implementation of disabled people's legal and constitution rights.

NOTES

1. 'Crip-politics' describes the attempt by people with disabilities to assert their differences by projecting positive images of themselves and their physical and/or mental impairments. In part, the movement mimics aspects of 'gay pride' and tries to play on pejorative stereotypes of disability by co-opting them and using them to highlight the demeaning ways in which societal values think and act about disability. In San Francisco, for example, 'crip' carnivals are held, while people with disabilities are actively involved in 'displaying themselves' at public events in order to raise the general level of public consciousness about their oppression and marginalization. In particular, the movement, while not homogeneous, tries to expose the hypocrisies of the mainstream, which, while apparently empathizing with people with disabilities, still incarcerates them.

2. I came across Ed Hall's paper long after I had written this chapter and only at the last moment did I incorporate some of his stimulating ideas. I would like to acknowledge his paper as one which makes a significant contribution to the debate about theorizing disability. To date it is the most innovative piece I have come across.

FURTHER READING

Oliver's (1990) book is one of the most comprehensive theoretical treatments of disability, although it suffers for its reductionist, materialist, framework. Birkenbach's (1993) text is one of the most considered philosophical accounts of disability and is to be recommended. Gleeson's (1995) article interlinks disability with geography and planning and represents a new departure in the literature, while Hall (1995) has produced a novel and important theorization of the 'body in context'. Susan Bordo's (1995) text on feminism and the body is a significant contribution in helping to deconstruct the ideologies of the body and embodiment.

3

State Policy and the (Re) Production of Disability

INTRODUCTION

We have said for two hundred years, and say anew each day, that we do not wish to see disabled people and disabilities. We do not want to be reminded of their needs, or burdened with their desires to live in the world . . . we prefer these people to be sequestered safely in secluded institutions – and they have been. We prefer them to remain second class citizens – and they are. We prefer them to be invisible – and they strive, many of them, to be so, in hope of perhaps that way finally of gaining our acceptance.

(Bowe, 1978, p. x).

The (re) production of disability in society is linked to a range of socio-institutional and political practices. In particular, a crucial context for the perpetuation of disablist attitudes and practices is in and through public policy and, as Oliver (1990) notes, the role of the state, in defining, categorizing and legislating for disabled people has been, and continues to be, a significant element in maintaining their marginal and dependent status. While state policies towards people with disabilities vary considerably between different countries, a range of common assumptions about the nature of disability, and of how to respond to it, is evident. In particular, an integrationist or assimilationist ethos – where the aim is to bring people back to 'normality' by creating the conditions for their integration into the mainstream of society – is increasingly supplanting the idea that people with disabilities should be segregated and hidden away from society. While, in part, such objectives are underpinned by the rhetoric of creating the conditions for disabled people to attain a state of independent living, the reality, as this chapter will document, is one where dependent ties are being retained, primarily through community-based services backed up by professional helpers.

In particular, a range of contradictions and paradoxes underpins the nature of state policies towards disabled people, not the least of which is that while an integrationist ethos is being espoused by many western governments, the reality for many people with disabilities, as chapter 1 has documented, is continued forms of socio-institutional segregation (Barnes, 1991; Dalley, 1992; Oliver, 1990). Moreover, the well documented restructuring of the welfare state, from public towards privatized forms of provision and care, combined with pressures being exerted by neo-liberal, conservative governments to reduce, in absolute terms, public expenditure, has brought the concept of utility to the forefront of social policy (Williams, 1992). This conception narrowly focuses on the costs and benefits of disability policies rather than measures of quality or of how people's lives are, or have been, enhanced. Indeed in the UK, unlike in, for example, Sweden and the USA, successive governments have resisted far-reaching civil rights legislation for disabled people, for fear of the apparent costs which might be placed on employers in having to comply with anti-discrimination legislation.

In this sense, the disempowered status of disabled people is clearly linked to ineffectual and weak statutory controls governing their access, and other, needs. As Oliver (1990) notes, in the UK context, the relevant statutes emphasize individual needs rather than social rights, and the Chronically Sick and Disabled Persons Act (CSDP) of 1970, still the most significant statute concerning access provision for disabled people, reinforces the idea that disability results from individual (personal) pathologies. In contrast, in the US, the Americans with Disabilities Act (1990) is underpinned by a framework which seems to provide a means for disabled people to challenge the structures of disablism, although, as Gilroy (1993) notes, it exhibits similar weaknesses as the existing, piecemeal, legislative base operating in the UK and elsewhere in that it is characterized by an unqualified voluntarism in only requiring access 'provided this is readily achievable'. Likewise, while countries like Sweden and the Netherlands have adopted strong statutes which assert the rights of people with disabilities, their underlying ideologies are, as Daunt (1991) notes, compliant rather than coercive.

Indeed, state policies towards disabled people, particularly in the UK and the USA, continue to be linked to the functional, and medically-derived, categories of the 'productive' and 'unproductive' worker, or, as Martin and Elliot (1992) argue, a characteristic of the state is a concern solely with identifying those who are regarded as 'legitimately' disabled (i.e. medically certified as such) to enable the selective provision of social security assistance and some dispensation with regard to work. Yet, as the chapter documents, the state's utilization of functional categories, to pigeon-hole disabled people, is also reinforced with forms of intervention largely typified by pluralism, voluntarism and market forces. Thus, as Williams (1992) notes, the rise of right-wing, neo-liberal, governments since the late 1970s, espousing an anti-state rhetoric of dissolving universal forms of provision has generated a 'mixed economy of welfare' characterized by the emergence of a range of fragmented, often privatized, providers of services to people with disabilities. Yet, as Oliver (1990) and others note, the emergent systems are still highly regulated by authoritarian central states, their resource bases are limited, while the evidence seems to indicate that profit and cost criteria, a balance sheet mentality, are the singular elements dictating standards of services.

In this chapter, I want to argue that the nature of state policies towards disability is, and has been, ill-suited towards the propagation of disability rights, and is implicated in maintaining and perpetuating disablism in the built environment and elsewhere (Morris, 1993; Oliver, 1990). In developing this proposition, I divide the chapter into two. The next section contextualizes the rise and development of state responses in Britain to disability by providing a critical overview of the transition from an ethos of segregation and punishment towards policies of integration and normalization, and from broad-based social welfare policies towards forms of market provision. In a second section, I trace the rise of market-driven, voluntaristic policy frameworks relating to disability and access by considering some of the similarities and differences between the UK and the USA in the development and implementation of access-related legislation. Indeed, while the civil status of people with disabilities appears to be markedly different between the two countries, the reality is one where disabled people in both countries are still struggling to gain strong and binding anti-discrimination legislation.

THE ORIGINS OF A PROBLEM: INSTITUTIONALIZING DISABLISM

The contemporary marginalization of disabled people is, in part, rooted in the breakdown of collectivized forms of production and the emergence of capitalist commodity relations (see Barnes, 1991). As the previous chapter indicated, the emergence of the wage relation, and the measurement of a person's value by virtue of their capacity to work, was a significant stage in the process of sorting out individuals into types. In particular, the physiological capacities of many people with disabilities were labelled a 'handicap' and disabled people were increasingly excluded from the major means of sustenance and social reproduction, employment. As a range of accounts has documented, the emergent factory systems were imposing new rhythms of work, new systems of order, or, as Ryan and Thomas (1980) note:

> the speed of factory work, the enforced discipline, the time keeping and production norms – all these were a highly unfavourable change from the slower, more self determined and flexible methods of work into which many handicapped (sic) people had been integrated (p. 101).

Indeed, as Barnes (1991) demonstrates, the new socio-economic conditions of the late eighteenth and early nineteenth century, with waged labour effectively making the distinction between the able-bodied and the disabled, generated new dependent populations which led to a specific response from the state, that of segregation. In the UK, for instance, the Poor Law Amendment Act (1834) was underpinned by a morality which not only individualized the nature of disability, but institutionalized state policies premised on the idea that people with disabilities should be disciplined, even punished, for their transgressions from normality (Schull, 1984). As Barnes (1991) has commented, 'segregating the poor into institutions . . . was efficient, it acted as a major deterrent to the able bodied malingerer, and it could instil good work habits into the inmates' (p. 15–16).

Such conceptions underpinned the emergence of new landscapes, built upon socio-spatial segregation, the assertion of a dominant morality based on a mixture of religious altruism and consciousness, and a curious Victorian patronage which, as Barnes notes, stemmed from the notion of the 'unfortunate cripple' and the need to help them with charity. Indeed, the development of charitable acts in the nineteenth century institutionalized what has since become a persistent feature of ableism, the projection of pity and empathy, of helping those who are conceived of as the 'unfortunate few'. Yet, charity was, and still is, subsumed by a wider segregationist ethos, of somehow keeping the 'diseased' and malevolent disabled apart form the normal population. By the end of the nineteenth century, a whole new pejorative language had emerged to describe people with disabilities, and the emergence, and persistence, of state-designated categories, like 'mentally sub-normal', 'the retarded' and 'defectives', became an established element of segregationist programmes which were premised on establishing forms of surveillance and control of those who were castigated as genetic and impaired deviants (Foucault, 1980).

By the turn of the nineteenth century, the segregation and exclusion of 'defectives' and the 'insane' had intensified because, as Hobsbawn (1968) notes, the so called 'second wave of industrialization', the development of heavy industries like iron and steel, was reinforcing the dependence of the economy on the 'fit and able-bodied'. Indeed, from this period onwards, the institutionalization and segregation of people with disabilities intensified and, as Barnes (1991) has documented, 'their numbers in special segregated places increased rapidly and did not begin to fall until the late 1950s' (p. 18). Throughout this period, the sustenance of segregation was, in part, linked to a range of (state sponsored) ideological justifications which portrayed the disabled as somehow a threat to the stability and functioning of society. As Barnes (1991) has documented, for instance, conceptions of social Darwinism, or the survival of the fittest, influenced late nineteenth century concerns that there were genetic links between physical and mental impairment and crime, unemployment, and other forms of social malaise. Indeed, the social eugenics movement of the time presented aspects of disability as a destabilizing phenomenon and was, in part, responsible for establishing compulsory sterilization programmes for mentally 'retarded' women in the USA.

While segregation as a form of social cleansing, even punishment, has remained an aspect of state responses to disability, the rise of the modern welfare state in the UK, post-1945, signified the emergence of paternalistic values and the propagation of the assumption that having a disability was synonymous with needing help and social support. Such paternalism reflects what Habermas (1975) refers to as the colonialization of the life world, where dominant groups project and assert the commonality of their ideas and values, seeking to inculcate and control 'others' through what Young (1990) refers to as the 'meshes of microauthority'. For disabled people, such meshes clearly relate to the rise of the welfare corporate state where a range of statutes and institutions asserted a form of state control over disability which was based on the notion of formal equality and impersonal procedures, that somehow the seemingly arbitrary and coercive forms of authority of a segregationist ethos were at odds with the emergent spirit of an enlightened welfarism. As Young (1990) comments, the new welfarism was seeking to deliver material gains and it seemed to hold some promise of an autonomy to groups like disabled people in a way previously denied to them through the authoritarian structures of segregation.

The emergent context, then, was evident in a range of legislation, from the Disabled Persons (Employment) Act of 1944, which tried to ensure jobs for disabled people, to the 1944 Education Act, which stated that all children regardless of abilities had to receive suitable education. In addition, the National Assistance Act of 1948 required local authorities to provide residential services and facilities for disabled people, so institutionalizing a new era of redistributive state provisions for people with disabilities. In particular, an assimilationist philosophy emerged which espoused integration into the mainstream as the 'cure' for disabled people, yet an integration which effectively sought to establish ableist rules and conduct through the extension of bureaucratic control. Indeed, the 'colonialization' of disabled people's life worlds has been endemic post second world war in seeking to deny their differences by imposing a universal conception of normalization upon them, of what it is to be a normal (able-bodied) person.

The resultant discourses of disability, then, tended to reflect a societal xenophobia, a collective neurosis about disability which fed off, and conditioned, the wider socio-institutional conditions within which most people with disabilities had to live their lives. In particular, the assimilationist ideal, or the notion that people with disabilities should be brought into the worlds of the ablebodied, revolved around the idea that people with disabilities should somehow seek to 'expel' their condition. Indeed, over the last twenty years or so, western welfare states have pursued strategies seeking to deny the differences of impairment, and the differences that impairments can make to the ways in which people who have them have to lead their lives. While this seems to signify a reversal of policies which previously segregated and excluded disabled people from society, people with disabilities are really being presented with a Faustian deal. It comprises the condition that the price to be paid by disabled people for their assimilation, of being accepted into the mainstream of society, is that they have to 'become normal', to somehow reject and deny their (bodily) differences, to, in Young's (1990) terms, 'scale their bodies'.

Indeed, the reality of assimilationist ideals, of their false promises, are everywhere for one to see.[1] In particular, divisions and differences persist precisely because people with disabilities are different and require contrasting services to those who do not share their physiological states. The experiences of assimilation in the USA, for example, denote the impossibilities, even the undesirability, of an 'integrated solution' if such solutions are premised on seeking solely to 'normalize' people with disabilities. For instance, prior to 1973 in the USA people with disabilities were not assimilated and segregated institutions were the norm. Much of the schooling system was less about education and more about medical care, control and, ultimately, gaining compliance from disabled people. As Dougan (1994), talking about his experiences in a segregated school, observed,

> I started at Sharpeville school and it was all about therapy not education. At the start of the day I was well ready to study but they'd take you out of your class for a 30 minute therapy session, drug you up, and you couldn't really function very well after that . . . I had therapy twice a day and could only ever concentrate on three classes out of six a day.

Even when Dougan was 'mainstreamed' and placed in a non segregated school, his physiological differences were identified as a marker requiring him to have 'special attention' from the school. A form of segregation still persisted in that, as Dougan said, 'I was always taught at one or two grades below what I was capable of and this really put me behind. In addition to this, they still sent me to special classes and really marked me off as different . . . I wasn't surprised that the other kids bullied me. I just looked different, I talked different, why should I have hidden it and why did the school want to get rid of those parts of me.' Assimilation, then, for Dougan, was more about the imposition of ableist values than about the propagation of sensitized, democratic policies, while his presence in a 'normal' school, far from de-emphasizing his differences (which was a stated objective of the school) accentuated them and left him open to the insensitivities of his fellow pupils (and even his carers).

For Dougan, his experiences were fashioned by a system underpinned by a dualistic structure which reinforced the dependent and marginal status of disabled people through the ideologies of expertism and clientism. In particular, the underlying rationality of the 'disability category' has been premised on rule by the professional expert, the all-knowledgeable, objective (hence value-neutral) bureaucrat with the alleged credentials of having attained the status of knowing about others and their life situations. For Giddens (1991) this is the pervasive characteristic of modern society, the rise of a meritocracy of professionals attaining control and power over others and exuding a sense that somehow they can be dispassionate while making the crucial decisions that make (and often break) people's lives. Moreover, the notion of clientism, or the formal submission of people to the rule of bureaucracy and of the expert, is a form of submission to the promise which the welfare state holds out, of redistributive equity. Yet, as Habermas (1987) notes, the documented history of welfare states indicates how officials 'not only prescribe much of the behaviour of clients or consumers within the institutions, but perhaps more importantly, through social scientific, managerial, or marketing disciplines, they define for the client or the consumer the very form and meaning of the needs the institutions aim to meet' (p. 362–63).

Thus, as Fine and Asch (1988a) and others argue, while the modern welfare state espoused the need to help, to provide support, it institutionalized forms of welfare dependence which, far from creating the conditions for independence and empowerment, did the reverse. Indeed, throughout the postwar period, as Dougan's (1994) testimony indicates, state policies retained a segregationist ethos in practice while increasingly espousing the rhetoric of 'normalization' and/or policies for integrating disabled people into the mainstream. Moreover, the role of the welfare state was largely confined to distributive issues of material well being and, as Young (1990) argues, the activities of the state largely sought to depoliticize the oppressive and domineering social relations that its clientele were ensnared within. In this sense, social regulation and control has, as Cohen and Roberts (1983) have argued, revolved around resolving 'who gets what' while smoke-screening questions of the organization and control of production and capital, the relevance of the practices and policies of both state and private institutions, and whether or not professional-client relations really represent a liberating means for people with disabilities to determine the ways they wish to lead their lives.

Lane's (1995) discussion of the (re) production of deafness provides apt illustrations of some of these points. She refers to the welfare state as comprising the 'troubled person's industries' documenting how deafness is 'constructed' by the state at the earliest age (of its clients) as a problem solely to be resolved through the 'administered benevolence' of the state apparatus. Thus, as Lane recounts, American audiologists have formally proposed testing the hearing of all American newborn, if only to expand their market (and with it the money to be made by selling their services). Those referred to an audiologist, as Lane notes, are labelled as 'a problem' as soon as they enter the testing laboratories, fitted out with all forms and types of equipment and made to feel as though their whole identity is solely that of 'their disability'. As Lane's study suggests, people who are hard-of-hearing and/or deaf are subjected to a plethora of experiments, tests, visits, re-visits, etc. but often with little or no justification. For Lane, and others, such situations indicate a certain corruptness within state welfarism and, as Lane argues, the professions 'stand the normal relation between needs and services on its head – services do not purely evolve to meet needs' (p. 174).

Indeed, the 'troubled person's industries' have expanded enormously since 1945 and, as Lane has documented, the professional services have fuelled the 'disability construction' of deafness by the perpetuation and expansion of their services and activities (p. 174). So, today, a veritable industry exists, experts in counselling, teachers, interpreters, audiologists, speech therapists, otologists, psychologists, and so on. As Lane says, quoting Gusfield's story about American missionaries who settled in Hawaii, 'they went to do good.

They stayed and did well' (p. 174). Moreover, the services provided are defined and delivered by individuals who rarely share the physical and/or mental impairment of the people who are receiving them, and Vaughan (1991) notes how visually impaired and/or blind people see professional service delivery as a dead end involving permanent dependence. Indeed, workshops for the blind are generally run by sighted managers while hard of hearing and/or deaf people are usually excluded from serving and/or servicing deaf individuals (because of their very impairment!). Thus, as Lane (1995) notes, one of the real, restrictive, ironies, is that deaf people are only able to qualify as teachers of deaf children after they have firstly qualified as teachers of hearing children, a route which, not surprisingly, poses considerable difficulties for them.

Yet, while the postwar period led to the rapid expansion of community based services and the proliferation of professional helpers the overall (assimilationist) approach never moved beyond the idea that somehow disability was a physiological condition requiring treatment, a cure, then rehabilitation. Indeed, as a range of researchers has noted, the whole ethos of the postwar responses of states towards disability was based on the idea that disability was somehow a static condition, that disabled people should, as chapter 1 intimated, adapt to hostile environments. Such assumptions, for instance, were the foundations of a range of statutes in the UK, from the Disabled Persons (Employment) Act of 1944 which provided rehabilitation services, to the statute which still governs access provision in the UK today, the Chronically Sick and Disabled Persons Act (1970). Likewise, in the USA, the development of a rehabilitation 'culture' emerged while a range of European countries was seeking to provide employment training and social rehabilitation for people with disabilities.[2] Yet, while western Europe as a whole had more or less accepted the need for positive discrimination for people with disabilities by the early 1960s, legislation was, and still is, weak, partial and increasingly being driven by the ethos of market utility.[3]

In particular, the transition to what some authors term a 'mixed economy of welfare' represents a broad based shift from the public to private sectors, from welfare to market forms of provision of care and support for people with disabilities. In doing so, the underlying ethos is premised on the idea that disabled people's needs can be best met, not by the provision of blanket universal services seemingly insensitive to the specific needs of individuals, but by the individuals themselves expressing their demands and 'purchasing' care from the relevant providers. Indeed, the language of the market is a key aspect of the emergent care regimes and, as Oliver (1990) notes, while, theoretically, the service providers should be sensitized to people with disabilities, they can only provide within their limited and fixed budgets. So, what disabled people are able to 'purchase' may not necessarily meet their fully expressed demands. Moreover, as Williams (1992) has observed, the attempt to break the power of the public sector and diversify the provision of care has, in many instances, done little to diversify the individual choices of disabled people. In particular, higher levels of uncertainty concerning funding has prompted the evolution of many more precarious, short-life, projects, while placing a premium on the efficient management of resources.

In part, this reflects the retention and dominance of hierarchical power structures between people with disabilities and those that provide for them. As Oliver (1990) comments, while the rhetoric and language of community care is about decentralism, consumer control and individual choice, the type of consumerism on offer is one which does little to put genuine control and choice at the disposal of disabled people precisely because the fiscal and resource base, and other forms of control, remain in the hands of the centralized authorities. Indeed, as Morris (1993) has noted, the focus on informal carers is really acting as a significant barrier to disabled people 'being considered as citizens in their own right' (p. 15). As Morris (1993) concludes,

> the words 'disabled people' rarely appear these days in policy documents and discussions without the words 'and their carers' tagged on behind. The meaning which is given to both the concept of 'caring' and of 'informed carer' is a crucial part of disabled people's experience of a lack of autonomy (p. 16)

Barnes (1991) interprets the present period of policy towards disabled people in the UK as a general retreat from the notion of rights 'towards voluntary rather than statutory based services' (p. 232, see chapter 9). As Oliver (1990) argues, in part, this is because the idea of giving civil rights to disabled people is anathema to governments which seek to propagate the rhetoric of serving special individual needs, as all that is required to 'solve' the problems of and for disabled people is to maintain their dependence on some form of servicing culture. Indeed, the identification of 'serving special needs' reveals the politically partial nature of policy and its unequal treatment of different people with disabilities. As Morris (1993) has argued, disability policy is still underpinned by the ideology of the efficacy of the functional and the productive body, of seeking to support those who are most able to move towards, and acquire some degree of, functional, work oriented, capacity. In this sense, those who benefit most from policy are not necessarily those that are most in need.

FROM CIVIL RIGHTS TO VOLUNTARISTIC CODES: A COMPARATIVE ASSESSMENT OF PLANNING FOR ACCESS

The wider ethos of public policy towards disability outlined above is, in part, reflected in the ways in which many western states approach issues of accessibility in the built environment. In particular, the dominant view revolves around the efficacy of the market in that, rather than being seen as citizens with full civil rights, people with disabilities, especially in the UK, are increasingly being presented as consumers in that if they want to exercise a particular demand for a commodity or a good (like an accessible building) then they are expected to pay for it. In the context of the built environment, a utilitarian view supports the idea that the provision of access facilities should only occur if a market demand can be expressed for them and rejects the view that developers should be made to provide additional features if little or no market demand exists. This (voluntaristic) conception, where developers are broadly left to decide as to whether or not they will provide specific forms of development, has been of increasing importance in spatial development since the early 1980s, yet it has generally worked against disabled access because of the perceived costs involved in either adapting existing buildings or providing 'special' features in new buildings.

In this sense, this perception of access as a commodity lies at the core of many planning approaches to disability and the built environment, where what is provided, its shape, size, and dimensions, is predominantly determined by the development industry (a theme I take up in more detail in chapters 5 and 6). Moreover, for people with disabilities, especially in the UK, the expression of their citizenship, or of their rights to an accessible environment, is increasingly being presented as the liberty of the individual to express choice in and through forms of market exchange, or a belief in the distributive justice of the market and of what Barton (1993) critically refers to as 'the centrality of choice within an enterprise culture' (p. 239). This, then, conjures up a concept of citizenship which reduces the disabled person to a consumer of goods and services, yet the dilemma for most people with disabilities is that their lowly and marginal status in the labour market, combined with systematic employer discrimination and a welfare system that provides minimal support, more or less consigns them to the realm of 'defective consumers' in that they have little or no income with which to exercise their 'citizenship' rights (Imrie, 1996b; Oliver, 1990).

Thus, there appear to be a number of paradoxes which surround the status of disabled people in relation to accessible environments, paradoxes which are perpetuated and (re) produced through the socio-institutional fabric of society. While, on the one hand, society seeks to assimilate people with disabilities into the 'mainstream', on the other hand, it identifies the mechanism of the process to be the market, or where the process of assimilation is reducible to commodity relations. Like all markets, of course, some consumers can engage, others cannot, so assimilation is necessarily an uneven, inequitable, process. Likewise, the centrality of bureaucratic control, underpinned by a technocratic ideology, is crucial in perpetuating disablism because, while the rhetoric of service provision has increasingly been one of 'responsiveness' to the consumer (if they can afford to pay), the reality is the persistence of the provision of universalistic services where varied policy users exist, trying to redefine people into categories which bear little relationship to them, their problems, or their daily experiences. Access, then, becomes something which is rarely defined in and through disabled people, yet, in acknowledging this, it is clear that there are cross national differences in policy practices.

In particular, I want to consider how such (utilitarian) values and related socio-institutional structures are the underpinnings of environmental planning responses to disablism in the built environment in both the USA and the UK. Yet, as the cases will illustrate, the ways in which such values are mediated by the specific socio-political formations of each country differ, and such differences, I shall argue, are crucial in the contrasting policy approaches to disability and access. Whereas the USA has a codified set of statutory regulations supported by the notion that access represents a key dimension of civil liberties, the UK is characterized by a weak legislative base which provides little or no basis for the legal control and/or enforcement of accessible environments. Indeed, a 'blame the victim' attitude pervades recent legislation towards people with disabilities in the UK and, as Barnes (1991) has noted, the government's attitudes towards the employment of disabled people, for instance, implies that employer discrimination can be justified. Thus, as the Department of Employment (1990) has stated:

> a major difficulty is that disability, unlike race or sex, can be relevant to job performance and what to some might seem like discrimination may in reality be recruitment based on legitimate preferences and likely performance (DoE, 1990 p. 30)

This is an astonishing statement yet wholly consistent with a succession of Conservative administrations since 1979 which have sought to minimize government regulations over the conduct of everyday life. Yet, in contrast, in the USA, statutory regulations and codes addressing access for disabled people have gradually shifted from broadly voluntaristic to mandatory forms of compliance. This reached its apogee in June 1990 with the signing of the Americans with Disability Act (ADA), by far and away the most wide-ranging and radical piece of legislation ever to be enacted anywhere concerning the civil rights of disabled people. Indeed, while in the USA individual civil rights are seen as the basis for pursuing accessible environments, and the absence of accessibility for a proportion of the population is deemed to be an infringement of a basic human liberty, in the UK there is little by way of an articulated (ethical) position. Successive British governments have refused to recognize that there is any infringement of human liberties, while continuing to exhort a voluntaristic pattern of provision of accessible environments. I compare and contrast these different positions and argue that mobility and access cannot be divorced from the (ethical) issues relating to civil rights and, ultimately, from questions of social justice.

(Re) Situating disability as a civil rights issue: the American experience

As a range of authors has claimed, the transformation in societal attitudes and behaviour towards disabled people is more marked in the USA than, arguably, in any other nation in the world (Bowe, 1978; Higgens, 1992). Indeed, the speed of change is evident if, for instance, one considers that in 1960 many municipalities still included in their local statutes what were termed 'ugly' laws, strictures which barred disabled people from public places on the grounds that their presence was offensive and posed undue legal liabilities. Historically, disability issues have been marginalized and generally ignored, and utilitarian ethics have tended to define the types and levels of provision for people with disabilities. However, since the early 1960s, a transformation in the political rights and social status of people with disabilities has been underway, a process which has been underpinned by the emergence of a radical disability lobby claiming that service and/or welfare provisions for disabled people is a civil rights issue or a matter of civil liberties. In addition, the more recent interlinking of disabled issues to wider concerns of economic and labour market efficiency and the restructuring of welfare provision has become a crucial means by which disabled people have been able to win the support of the political, pro-market, pro business mainstream, a support which is reflected in the wide-ranging statutory provisions which now oversee access and other disability issues (see Figure 3.1).

Indeed, as Myette (1994) notes, a crucial part of the process was the emergence of civil rights campaigning and especially the power of the

Figure 3.1   The gains which various people with disabilities have achieved are partly indicated by the person in a wheelchair waiting for a bus to pick him up just below Capitol Hill in Washington D.C. The majority of buses in Washington D.C. are wheelchair accessible so this is a common sight around the city. Yet, it would be remarkable to see a similar situation in the United Kingdom.

articulate, educated, Vietnam war veterans, people who had been 'able-bodied' and then, suddenly, acquired an impairment. As a range of observers has noted, their increasing dissatisfaction with the range of voluntary codes, standards and agreements concerning access was crucial, and throughout the 1960s frustration with the absence of monitoring systems to ensure compliance with access standards was increasingly being expressed through a mixture of both indirect and direct action (see Bowe, 1978). At the congressional level, disability issues were also being debated and by the early 1960s the first wave of federal government recognition of disability as a form of social and political marginalization was being expressed. In a congressional hearing, for instance, a representative commented that, 'the status of handicapped (*sic*) people in this country is a disgrace and it has no place in a decent, civilized, society . . . we must do all we can to deliver these people back to the nation' (Congressional Record 110: 3138, 1964).

Such sentiments, while paternalistic, reflected an increasing awareness in the USA of the second class status of disabled people and, in part, precipitated the first wave of 'barrier-freedom' legislation. The important elements included the introduction of national standards for barrier removal in 1961, through the auspices of the American National Standards Institute, the Architectural Barriers Act of 1968, and the Rehabilitation Act (1973), a statutory element which attempted to reject the idea that rights should only be protected in segregated, separate and special environments. This was a break-through inasmuch as it declared that disabled people had 'equal rights' and should be given 'fair and equal treatment'. It introduced client assistance programs and advocates to represent disabled people in claiming 'their rights', while state rehabilitation agencies were required to involve disabled people in the development of their annual plans submitted to federal government. Thus, the status of disabled people seemed to be enhanced yet, as Bowe (1978) and others have noted, the measures were rarely implemented in the spirit they were intended.

In part, this underpinned the emergence of radical disability pressure groups increasingly critical of assimilationist ideals and of the paternalism vested in the legislation. Thus, from the early 1970s onwards, a range of groups asserted the specificity of 'disability culture', of its internal heterogeneity, its vitality, and of the need for people with disabilities to be empowered, rather than being absorbed, by state welfare institutions. A 'crip-politics', celebrating the diversity and difference of disability, has since emerged, while the state has been forced to redefine and expand its conceptions of what disability actually is. Such pressure has been exerted, in part, by cross class alliances and coalitions among a range of disparate pressure groups. Coalitions, for instance, have developed between the AIDS lobby, women's organizations and 'grey power', while, more recently, their lobbying for civil rights legislation for disability has sought to interlink a radical, left, politics, with the social and political concerns of the right wing in America, a manoeuvre described by Myette (1994) as, 'a clever way of coopting national support . . . because what we've seen is disabled people asserting their differences and independence from the controls of state welfarism and so appealing to the Republicans' desire to reduce welfare expenditures, all of which was crucial in the successful passage of the ADA'.

Scott (1993) concurs with this in noting that the disability lobby 'did not present itself as a threat to the American Dream . . . but portrayed disability issues as an intrinsic and vital part of a society based on individual contributions and wealth based on the potential of its citizens' (p. 4). In this sense, the appeal was to specific utilitarian concerns, the essence of individual libertarianism, for choice, and the rights of all to be free and equal citizens. Moreover, such notions were increasingly being aligned to what Gooding (1994) has characterized as the 'key policy goals like labour market efficiency and curbing growth of welfare dependency' (p. 5). By the mid 1980s, such alignments between disability rights and wider policy goals were being espoused by leading politicians and their advisers and, as Evan Kemp, the Republican leader of the Equal Employment Opportunities Commission (1990), commented:

> There are good dollar and cents reasons why business should be interested in disabled persons. First, disabled people purchase goods and services just like any other consumer . . . A smart business person would make sure that his or her business was accessible to and usable by disabled people . . . 36 million Americans . . . can be a profitable market for you (p. 2).

Likewise, in 1990, President George Bush appealed to narrowly focused utilitarian concerns while linking disability issues to wider welfare agendas. As Bush commented, 'when you add together state, local, and private funds it costs almost $1200 billion annually to support Americans with disabilities, in effect to keep them dependent'. Such observations were also reiterated at congressional hearings into the proposed ADA. Indeed, as Senator Edward Kennedy outlined, 'some will argue that it costs too much to implement this bill. But I reply, it costs too much to go on without it. Four per cent of American gross domestic product is spent on keeping disabled people dependent.' In this sense, disability issues were being presented as less about the autonomy and civil rights of the individual disabled person and more about restoring the federal reserves while reducing the tax burden generated by the welfare demands of people with disabilities.

The subsequent passing of the ADA, then, seemed to represent the apogee of the disability civil rights movement by seeking to eliminate all forms of societal discrimination against disabled people. Its provisions are wide-ranging and they include anti-discrimination measures which permit the federal government to initiate litigation against transgressions of the statute. The definitional basis of disability covers every form of impairment that can possibly be identified, including people with AIDS. In particular, barrier-freedom in all new public buildings is specified as compulsory while any renovations and/or redevelopments of existing buildings require access to be provided for all categories of disabled people. Unlike other civil rights legislation, the ADA also permits individuals to file complaints of transgression to federal district courts, while injunctions can be served prior to the construction of a building if an individual complains that they are about 'to be discriminated against'. The evolving system of planning for access, then, is hierarchical and top-down with regulations and codes being 'handed out' by the federal government.

The ADA also specifies in detail the variety of access requirements, from the dimensions of doorways to the elevation of walkways. At the local, city, level, planners are expected to utilize the ADA as the basis for decisions as to whether or not to issue planning permits and, as the chief permits officer in Berkeley, California, commented, 'every city in the USA needs to identify structural barriers to access and develop transition plans to remove them'. The implications of the ADA, then, seem potentially far reaching in requiring municipalities to develop such plans or schemes which identify which buildings are inaccessible with suggestions as to how such inaccessibility will be overcome. The ADA also prohibits any public activity being carried out in inaccessible places and, as Myette (1994), head of Architectural and Planning Compliance in Boston, indicated, 'if we hold any fetes or fairs they must all be accessible and we've now listed all the public buildings and what their accessible status is . . . if there are any barriers we'll remove them or the activities from there'.

Yet, as Higgens (1992) notes, while the ADA, at face value, is strongly 'rights based', its essential character is conservative while wedded to a market welfare ideology. Thus, the ADA permits 'get out' clauses like 'undue hardship' and 'readily achievable' to be used in deciding whether or not commercial operators have to remove barriers or facilitate better levels of access to their buildings. Employers are also exempt from the ADA when to meet its provisions would entail 'unreasonable financial costs' or where they have fewer than 15 employees. Moreover, domestic dwellings remain beyond the strictures of the legislation, while operators of public facilities, like transit authorities, have thirty years to make existing stock accessible. Indeed, as Higgens (1992) has documented, many corporations in the US are resisting the ADA and, in 1993, the chairman of the Metropolitan Transit Authority of New York claimed that, 'it'll cost us $100 million to make the subway accessible but to do so would benefit few because few disabled people would use the subway'.

While a utilitarian ethos of this type is seeking to dilute access controls it also appears that the 'rights' approach to access policy is problematical and suffers from limitations which render it less than applicable to the British, or any other, context. As Higgens (1992) notes, it reinforces conceptions of disability as somehow an individual problem, that is, the retention of "a general presumption that differences reside in the different persons rather than in relation to norms embedded in prevailing institutions – it also presumes that the status quo is natural and good' (p. 230). It also reinforces a form of legal individualism, so ignoring, even denying, the role of (collective) structural inequalities in perpetuating discrimination against disabled people. As Cotterell (1992) has argued:

> the individualism embodied in modern law stresses above all that individuals are makers of their own destiny; standing alone they bear responsibility for the omissions attributed to them . . . in its purest forms, it takes no account of social or cultural factors (p. 119).

Moreover, Gooding (1994) notes that rights legislation tends to offer equal protection to manifestly unequal groups in society, with little sense that positive discrimination might be the only option to turn around structural disadvantages. As Gooding (1994) comments, the underlying basis of civil rights legislation, particular the sex and race laws, is its premise 'upon the existence of a neutrally functioning, non discriminating society where only isolated acts of irrational discrimination occur' (p. 31).

Likewise, Young (1990) argues that part of the difficulty with the rights approach is its conception that the problem of and for oppressed people is one of discrimination because 'identifying group-based injustice with discrimination tends to put the onus on the victims to prove a harm is done, case by case' (p. 71). For instance, one of the clauses of the ADA states that enforcement 'relies on individuals', that they 'can bring an action if s/he is about to be discriminated against'. Yet, as Young and others note, discrimination as a concept is problematical because it presents injustice as an aberration, a temporary phenomenon, that given its absence then all injustices would disappear. However, this begs the question of whether or not injustice can be reduced to the notion of discrimination and/or whether or not discrimination, as a concept, necessarily defines all forms of societal injustices. In this sense, the rights approach trivializes disability and fails to develop a critique of 'normal society, of its sociocultural practices and institutions, or of its oppressive social policies'. As Higgens (1992) concludes, a rights rationale tends to say that 'if you measure up, then you have rights – the rights that those of us who are non disabled have enjoyed – and may extend some special treatment but that's what it is' (p. 231).

Yet, as Smart (1989) has argued, the real strength of the rights movement is that to 'couch a claim in terms of rights is a major step towards a recognition of a social wrong' (p. 143). Indeed, a range of authors concurs that the tendency to portray, for example, disability allowances as a government handout and/or benefit reflects the idea that the disabled person should be grateful for what they receive, an idea which is undermined by a rights discourse. In distinction, a rights discourse is premised on a critique of charity and of state benevolence or welfarism. In this sense, such discourses are potentially empowering and transformative by positioning disabled people's subordination as a product of structural socio-political disadvantage. As Williams (1987) has noted:

For the historically disempowered, the conferring of rights is symbolic of all the denied aspects of humanity: rights imply a respect which places one within the referential range of self and others, which elevates one's station from human body to human being (p. 416).

This, then, is a critical positioning for people with disabilities. Planners in practice are obliged to be attentive to access in ways which transcend access issues being conceived as technical and/or design issues but as something which is clearly linked to issues of social (in) justice. In this sense, the ADA, and related barrier-free legislation have attempted to interlink disability and access to moral and/or ethical questions which, while seeking to off-set the underlying utilitarianism of American society, really goes to the root of the question of 'how should people be expected to live'? As Dougan (1994) has commented, 'whatever all its limitations, at least the ADA and the general awareness about access and disability is helping us to show how the built environment has, and still continues to, oppress us . . . we're now firmly on the political fixture list', an observation which is well removed from what we can observe in the British context.

Public policy and access: reinforcing voluntarism in the United Kingdom

In 1979 the Snowdon Commission, or Silver Jubilee Committee, in investigating access for disabled people in Britain, concluded that the general inaccessibility of most parts of the built environment was tantamount to an infringement of the civil liberties of people with disabilities. Yet, throughout the 1980s, while access as an issue gathered momentum, and even some statutory backing, little was achieved. Indeed, while successive governments have acknowledged the discriminatory nature of much of the built environment, regulatory provisions are still circumscribed by a government which, at the time of writing, is obsessed with the pursuit of utilitarian ideals. The government's position on access seems clear and a recent statement by David Curry (1993), Minister for Local Government, reinforced the centrality of costs and market utility in that:

whilst committed to creating an environment more accessible to people with disabilities we must ensure that any additional costs do not bear unreasonably heavy on those who provide and use buildings or on the community which ultimately pays the price for goods and service.

What seems so astonishing about this statement is the way in which it (apparently) excludes people with disabilities from the categories of 'user' or 'community' as though they somehow are external to society, that they do not, in themselves, pay in some way for the form and structure of the built environment.

The contrast with the passage of the ADA is stark. In the USA, people with disabilities were placed at the fulcrum of the underlying definition of community: disabled people were conceived of as full citizens with equal rights. Yet, in distinction, the UK context has elevated the requirements of the development industry which are seen as paramount in government thinking on access and as Allberry (1992), private secretary to the Minister for Housing and Planning, commented to the RTPI:

> we have promised a further circular on access considerations, but the issue of this will need to await, first, some very necessary research defining the requirements, cost, and practicality of improving access standards and secondly consultation with the public and the industry as the best way of meeting these objectives.

What we see here, then, is a refusal to recognize the legitimate (mobility and access) rights of a substantial proportion of the population and the minister's attempt to decouple the category of 'the disabled' from 'the public and the industry', somehow seeing the former as separate and divisible from the latter. As Barnes concludes, 'for the foreseeable future the majority of disabled people will still not be able to go where able-bodied people go' (1991, p. 179).

This reflects a combination of wider, structural, processes, not the least of which has been the Conservative government's political and economic decontrols of the last decade, where meeting the needs of disabled people is increasingly being identified with forms of market provision, as distinct from regarding the demands of disabled people as constituting a 'right'. This reflects the wider neo-liberal philosophy that has been popularized in Britain, a series of ideas which has also served to curtail the fiscal and legal powers of the dominant service providers, the local authorities. Indeed, the centralization of political power in Britain since 1980 reflects the government's desire to restructure the way 'bureaucratic institutions affect our industrial and economic life' (HMSO, 1989) by placing markets at the fulcrum of service provision. Yet, as Townsend (1979) and others have shown, many disabled people do not possess the economic resources to create an effective demand for the market provision of accessible environments, while the diminishing resource base of local authorities, coupled with the ineffectual political power of disabled groups, has done little to increase the public provision of more accessible built environments (see Imrie and Wells, 1992).

Likewise, the continuing political dispute about disability rights in the UK is focused on the perceived costs any rights-based legislation would impose on employers. Thus, the failure of Roger Berry's Civil Rights (Disabled Persons) Bill, introduced to the House of Commons for a special reading in May 1994, to gain government support, despite the government's protestations that they support disabled people's rights, reflects the wider political pre-occupation with de-regulating labour markets and minimizing costs. Such concerns were reinforced in early 1995 with the introduction of the Conservative government's own version of a disability rights bill, the Disability Discrimination Act, which, while making it illegal for employers to discriminate against disabled people, provides exemptions for firms who employ fewer than twenty people. As Hague, then Social Security Minister, commented, 'we do have to recognize that there will be a cost for many employers. We expect that the cost in the case of goods and services providers to run into hundreds of millions of pounds' (quoted in Goodwin, 1995, p. 5).

The underlying voluntarism is particularly evident with the Disability Discrimination Act which, while insisting on certain rights of access to public buildings and places of employment, does so in an ambiguous and ill defined way. For instance, cinemas, restaurants and shops are only required to provide access 'where reasonable' and, as Laurance (1995) notes, a restaurant could, for instance, have only a single designated table for a disabled person while a corner shop would be within the law if it provided a bell outside which a disabled person could ring. In addition, access to the London underground system, and to what the bill terms 'similar transport systems in other cities', will only be required where new stations are being constructed. As Bradshaw (1995), disability campaigner, has commented, 'the bill is a set of half measures which were neither comprehensive nor legally enforceable . . . many employers will remain free to exclude and discriminate against disabled people . . . there is no central commission which would enforce our civil rights and we will have no legal aid for redress to tribunals' (p. 7). In this sense, the voluntaristic ethic is preserved and the concept of utility left as the centrepiece of disability policy.

Such voluntarism is also at the heart of the statutory basis of the British planning system and the range of devices available to planners to enforce access is heavily circumscribed by wider utilitarian concerns. Throughout the 1980s, directives never made access compulsory in the way, for instance, fire regulations are and by the mid 1990s access is still underpinned by the Chronically Sick and Disabled Persons Act (CSDP, 1970) which requires developers to provide access only where 'practical and reasonable'. While Part M of the Building Regulations, introduced in 1987 and extended in 1992, provides further guidance on access, it refers only to new public buildings while excluding domestic dwellings and most minor building renovations and redevelopments (see chapter 5). In addition, government directives and circulars have made it clear that prescriptive policies on access in local plans will be deleted while few, if any, local planning authorities really regard the setting of planning conditions, as a means of securing access, as 'legitimate' (see Imrie and Wells, 1993). As Imrie's (1996b, 1996c) research indicates, local planners can be little more than reactive given the absence of clear statutory backing with which to pursue access (see chapter 6).

In particular, research by Imrie (1996c), and Imrie and Wells (1993), on access and planning indicates how access issues remain marginal to the main work of planning departments. This work indicates that most planners in practice operate with a definition of disability which is partial, primarily defined in terms of 'people in wheelchairs'. Moreover, very few authorities have any knowledge of the (local) demand for accessible environments (or they do not know who 'the disabled' are). It is also common for planners to link access issues to the market opportunities afforded to developers or, as one local planning authority noted, 'our attitude to the issue [of access] is to tell the developer that an accessible building is one which maximizes their market opportunities' (quoted in Imrie and Wells, 1993, p. 217). The coupling of access with market opportunities is an obvious reflection of the underlying utilitarianism driving British public policies yet, as one planner commented, 'unless we can demonstrate to developers that there's money to be had from it they just won't bother' (quoted in Imrie, 1996c). While access issues are undoubtedly higher up the policy agenda in UK local authorities than was the case five years ago, the overall conclusions still suggest that access issues are a low priority, attracting few funds or resources, with few authorities seeking to develop separate budgets to manage or develop access policies (see Imrie and Wells, 1993).

Such situations are related, in part, to the weak and marginal political status of people with disabilities in the UK. Disabled groups have yet to clearly articulate a politics of difference or to assert their self defined agenda which is not part of integrationist policies (see chapter 7). In particular, the politics of access has been constructed around a range of assimilationist ideals defined, largely, in and through the state and with the broad objectives of, in Young's (1990) terms, denying that group differences, like a disability, 'can be positive and desirable' (p. 166). Indeed, the charitable ethos, which continues to underpin much of disability policy and politics in the UK, reinforces the notion that people with disabilities should be grateful for what they receive and, until recently, the splintered, non radical, nature of disability pressure groups did little to challenge the paternalism of the state. This is especially evident in relation to access issues whereby a system of advocates, or access officers, in local authorities has emerged since the early 1980s, primarily with the objective of representing disabled people's access concerns within local formal political systems. Yet, as Imrie (1996b) has outlined, the institutionalization of access in this way seems to be doing little more than perpetuating a professional-client relationship, somehow reinforcing expertism and clientism and the idea that people with disabilities are little more than passive consumers of the services of the welfare state.

Indeed, the politics of the type that the systems of access seek to propagate has the effect, as Young (1990) argues, of locking 'individual citizens out of direct participation in decision making , and often keeps them ignorant of the proposals deliberated and the decisions made' (p. 73). For the most part, it is a pluralist politics of a type which fails to recognize the 'individual' person and people are only permitted to voice their concerns through the context of specific government programs and/or formally constituted institutions. In this sense, there is an effective decoupling process or one in which, as Young describes it, a particular individual's self interest becomes 'incoherent' (p. 73). In the context of the access officers' multiple roles, it is clear that their dealings are wholly defined in terms of (an idealized) 'interest group pluralism' and, as one access officer described in interview, 'we'll only deal with the recognized groups, we haven't got time to do any more than this and anyway this is the only way to get anything done'. Such, then, is the British system of policy pragmatism for 'dealing with' access and disability, a system seemingly without any real, articulated, moral basis and resistant to the possibilities of the societal oppression of disabled people by virtue of the practices of the dominant socio-institutional structures.

CONCLUSIONS

The underlying rationale of a disability policy, in both the USA and the UK, has, historically, been driven by utilitarian concepts of costs and benefits, while predominantly focusing on the individual pathologies of disabled people rather than on the socio-institutional prejudices and barriers that they have to confront as an everyday experience. In both countries, the structures of the state remain broadly wedded to disablist values while exhorting disabled people to conform to their environment. However, significant variations exist between the two countries. While disability and access issues in the USA, for example, are articulated as matters of social (in) justice, a problem of and for civil liberties, in the UK access is seen, by government, as a technical and/or compensatory matter which can be dealt with through the appropriate redistributive measures. Likewise, while planners in the UK have to utilize a weak regulatory framework for pursuing access issues with developers, planners in the USA are supported by a much stronger framework for securing access provisions. In addition, access in the USA is underpinned by a widespread politicization of disabled people while, in contrast, disabled pressure groups in the UK, although increasingly vocal, remain weak, splintered, and reactive (see Barnes, 1991; Oliver, 1990).

In a wide ranging review, Oliver (1990) notes how the development of a disability politics in the USA pre-figures the increasingly dominant view that disability is not merely socially constructed, but socially created as a form of institutionalized social oppression, like institutionalized racism or sexism (see Oliver, 1990, p. 121). Interestingly, while a radical disability politics in the USA has emerged as a response to the perceived failure of existing political institutions and strategies to represent disability as 'a right', the movement in the UK has been more circumspect and, so some would argue, depoliticized. Indeed, while movements in the US and elsewhere seek to place the position of disabled people within 'structures of oppression', British disability groups still tend to concede a need to conform to 'normal' society (see Barnes, 1991; Barton, 1989; Oliver, 1990). In part, the US situation is linked to the ways in which civil liberties and rights are enshrined in the constitution, whereas, in the UK, it is still perfectly legal to discriminate against a person on the grounds of their physical and/or mental impairment, a quite astonishing situation.

Such contrasts, however, tend to mask the similarities between the two countries and their shared conception that somehow people with disabilities should be subjected to the strictures of their respective welfare states, while undergoing a broad-based assimilation into the 'norms' of society (Bordo,

1995; Imrie, 1996b; Laws, 1994a). Yet, following Young (1990), assimilationist strategies which try to eliminate group and/or individual differences must be resisted precisely because they seek to propagate a 'blame the victim' approach to oppression while asserting the primacy of equal treatment as the 'principle of justice'. Yet, equal treatment in itself is precisely what many people with disabilities do not require in that their physiological differences, their physical and/or mental impairments, often require differential or unequal treatment (how else can it be addressed?). Indeed, a radical politics of access is more or less saying that the key principle of distributive justice in western society, that of 'fair and equal treatment', is failing to grasp the reality of the differences which people with disabilities exude. That is, for them to participate in society really requires something which is much more sensitized to their individual needs, revolving around a celebration, not a castigation, of their differences. This, then, is one of the key challenges.

NOTES

1. One of the most problematical forms of segregation and/or exclusion occurs in housing and research indicates that housing organizations continue to pay little or no attention to equal opportunities or to the specific requirements of groups like disabled people. As Laune (1993) has argued, 'the practice of ghettoizing disabled people into the category of special needs allows mainstream housing provision to ignore the needs of disabled people and confirms images of the deserving poor' (p. 2). The difficulties facing disabled people's access to housing are varied and complex yet the underlying situation clearly indicates that the situation has worsened over the last ten years. Indeed, as Laune comments, 'all of the available information on disability, access, and housing shows that far from facilitating personal autonomy, independent living, and equality of opportunity, the reverse has happened' (p. 3). Indeed, the number of local authority and housing association wheelchair and mobility standard property housing starts fell significantly in the 1980s. Thus, as Laune has documented, while in 1979 housing associations built 129 new wheelchair standard houses, in 1990 the figure was 67. Likewise, local authorities built 576 such properties in 1979 but only 69 in 1990. In the respective years, the numbers of new mobility standard properties fell from 2136 to 102 in housing associations and from 5950 to 469 in local authorities.

2. In France in 1957, for example, the first statute on vocational and social rehabilitation appeared, while in Belgium, in 1963, and in Italy, in 1968, regulations for obligatory employment were drafted into law.

3. As Morris (1993) shows, since 1988 the new system of community care for people with disabilities brings the rhetoric of the market into the system of care by treating people with disabilities as consumers who, in being assessed by care managers, are dependent upon them for purchasing the services and/or goods. Indeed, the perpetuation of a 'culture of dependence' is evident in the Disabled Persons Act (1986) which, while seeking to involve disabled people much more in shaping the services that they receive, provides little guidance on how their rights to be 'assessed, consulted, and represented' can be attained in the face of, for example, hostile and non cooperative local authorities. For Oliver (1990), far from the restructuring of the welfare state

diminishing dependence, it seems, potentially, to be extending the ethos of professional and administrative approaches to the problems of disability.

FURTHER READING

Barnes (1991) provides a general overview of the changing interrelationships between the state and people with disabilities. One of the best reads on the interrelationships between state policy and disability is Stone (1984), who interlinks the category 'disability' with the emergence of industrial capitalism. Gooding (1994) provides a wonderful exposition of disability, law, and the state in America and Britain, while Scott's (1994) account of the Americans with Disabilities Act provides a useful overview of the situation facing people with disabilities in the USA.

4

Designing Disabling Environments

INTRODUCTION

As materials for culture, the stones of the modern city seem badly laid by planners and architects, in that the shopping mall, the parking lot, the apartment house elevator do not suggest in their form the complexities of how people might live. What were once the experiences of places appear now as floating mental operations.

Sennett, 1990, p. xi.

Sennett's powerful statement about the emergence of the modern city depicts a dilemma of contemporary urbanism, that is, the production of spaces which discourage social interactions and the neutralization of public space in the context of an overarching privatization of the built environment (Giddens, 1991; Harvey, 1990; Laws, 1994a, 1994b). For Sennett and other commentators the socio-cultural problem of the modern city is how to make the impersonal milieu speak, how to relieve its current blandness, its neutrality, while breaking down the social and physical barriers which segregate, seclude and, ultimately, reinforce discrimination and disadvantage against many of its citizens (also see Dickens, 1980; Imrie, 1996b; Jacobs, 1961; Jencks, 1987). Indeed, as a range of authors concur, of particular concern is how the built environment both reflects and conditions wider processes of socio-economic change, of the intersection between global corporate capital, its imperatives for particular spatial configurations, and the facilitative roles of the agents and institutions of change in the built environment, especially of architects, building engineers, and town planners (Knesl, 1984; Knox, 1987; Ward, 1994).

As Knox (1987) has noted, while there is extended documentation of the new urban economies and of the interrelationships between city structures and wider global processes, relatively little attention has been paid to 'the agents and outcomes of change in the built environment' (p. 354–55). Where discussion does exist, it has either tended to trivialize the role of the architect as passive, as an instrument of the client, or elevated them to a position of supreme control, able to fashion the built environment as though it were purely a product of design. Yet, as Knox and others have noted, neither conception is adequate because they, in Knox's terms, either understate 'the broader context of social and economic forces (as modulated and amplified by institutions) or overplay the roles of architects' (p. 355, also see Knesl, 1984). Yet, clearly, architects, and other design professionals, are implicated in the production of the built environment, of developing aesthetic values, propagating specific conceptions of design, and engaging with wider social structures. In this sense, the socio-economic, political and, crucially, ideological relations of architectural theories and practices are of vital importance to explore in order to gain some understanding of how the uneven, and unequal, spaces of the built environment are developed and perpetuated.

In discussing such themes I divide the chapter into four. First, I consider the interrelationships between architects, power and the built environment, where I develop the argument that the perpetuation of disablist spaces is critically linked to the socio-institutional practices of architects and the wider design professions. Second, I then relate such ideas to the importance of modernism in the construction of the disablist city. While modernism, as a set of ideas and related socio-political practices, is not exclusively responsible for the construction of disablist cities, it can be argued that it has been the dominant force in their postwar reconstruction (Harvey, 1990; Giddens, 1991; Knox, 1987). In a third section, I contest the notion that the development of a post-modern architecture signals the possibilities for a liberating, non oppressive, built environment. While, theoretically, elements of post-modernist ideas and practices provide for the possibilities of an engagement with subjectivity, of creating the conditions for a voice for people with disabilities, there are clear concerns about whether or not modernism has been transcended and also the extent to which a post-modern politics (if such a thing can ever exist) has the capacity to create non disablist environments (see O'Neil, 1995; Taylor-Gooby, 1994). In a concluding section, I consider the relevance of principles of universal design and of the possibilities of an emancipatory architecture.

ARCHITECTURE, POWER, DISABLISM: THE CONNECTIONS

Over the last twenty years a powerful critique of the role of the architect, in the perpetuation of gendered, racial and other divisions in the city, has emerged (Dickens, 1980; Knox, 1987). It is based on the idea that the interplay between the ideologies and institutional practices of the design professions, within the wider context of particular socio-economic strictures, has served to exclude minority interests while reinforcing an alienating and oppressive built environment. The documentary material, as chapter 1 indicates, ranges widely from accounts which show how the built form is inattentive to the needs of women to those which suggest that spaces are segregated on a racialized and disablist basis (see Thomas, 1995; Golledge, 1993; Imrie and Wells, 1993). Indeed, as Matrix (1984) notes, there is an assumption by architects of 'sameness', of normality, amongst the population, 'that all sections of the community want the environment to do the same things for them' (p. 3). Such ideas have been sustained through three interconnected dimensions of the design process, that is, the (ideological) assertion of the aesthetic or prioritizing the idea of building form over use, the professionalization of architectural and other design practices, thus creating a new technical, 'expert', elite, and the rise of the corporate economy as the dominant clientele.

In considering the relationship between aesthetic values and the production of the built environment, Ghirardo (1991) has noted how many architects still see their practices as about the designer providing buildings with critical capacities, that the architect can engage 'with contemporary problems through formal manipulation' (p. 12). In such views, it is assumed that architecture is a form of artistic expression and endeavour and, in Ghirardo's terms, 'that art has a high moral purpose in the formation and transmission of culture . . . of the design of aesthetically pleasing forms of poetic spaces' (p. 9). This, then, projects the architect as a purveyor of beauty and truth, an elevated being somehow with the abilities and skills to construct for (as distinct from with) the population as a whole (Dickens, 1980; Knesl, 1984). Indeed, such conceptions have performed a powerful ideological role in architecture, especially, as Porphyrios (1985) has argued, one of self legitimation through the perpetuation of discourses which seek to elevate the practices of architects to a form of objective neutrality, the idea of the rational technicist operating for a willing (and compliant) clientele (also see Sennett, 1990; Wolfe, 1981).

In particular, as McGlynn and Murrain (1994) note, it has never been a feature of the culture, social ethics and/or practices of design professionals to see themselves as part of wider political processes. As they comment, architects seem to have limited understanding of the relationships between values, design objectives, and the design intentions derived from them, with design theory tending to concentrate on the technocratic and technological, reducing questions of access and form to the functional aspects of the subject, yet ignoring what Davies and Lifchez (1987) have termed the social psychology of design or trying to understand what it is that people really want (also see Dickens, 1980, p. 353). In this sense, as Davies and Lifchez (1987) have argued, the popularization of architecture as 'high art', or pure design, is underpinned by a capacity to perpetuate an impersonal, often alienating, practice, given that the focus is about the aesthetic, or the building form, not the user and/or the pragmatics of the functioning of the building. Buildings, then, in this interpretation, are treated as an abstraction, something over and beyond, somehow able to transcend, the socio-political contexts within which they are produced (King, 1984; Schull, 1984).

Such conceptions, as the next section of the chapter will show, reached their apogee under the postwar modern movement where the emphasis on minimalist form sought to reduce the complexity of human movement and building use to a singular set of rules and/or laws, or the idea that all human action is knowable and controllable. As Sennett (1990) and others have commented, the idea of control, coupled with the perpetuation of the ideology of architect as artist, was simultaneously disarming and disabling in a number of interrelated ways. Foremost, it perpetuated a representation of the architect as 'expert', so providing a legitimation to practice unfettered by wider public and/or corporate controls. In this sense, the architect was more or less untouchable (Wolfe, 1981). In addition, the 'expert' characterization, reinforced by architects aligning themselves to the idea that their practices were somehow supported by a scientific rationalism, was crucial in signalling to a wider public that they were there to be 'acted on', that architectural knowledge was something to be handed down, or a form of received wisdom. Such paternalism was, and still is, a crucial ingredient in denying the subjectivities of the very users of the built environment.

Yet such paternalism has been vigorously defended and a combination of institutional indifference, and resistance to conceptions which undercut the coherence of the idea of 'architect as artist' and/or rational technicist, is evident in a range of contexts, not the least of which is where designers seek to propagate their professionalism. For instance, in one of the more progressive cities to propagate disabled people's rights, San Francisco, there is documented resistance by architects towards what they see as the dilution of the aesthetic integrity of their designs, by virtue of being instructed by the city authorities to 'moderate all designs by following the requisite access regulations' (Access Code for California, 1993) . In interview, the access officer for the city commented:

> I try and train the city architects, I run regular courses, but they're not very receptive, they don't want to spend time on it . . . universal design isn't important to them; they see access as a separate issue, as an additional design requirement which they think just compromises what they're trying to do . . . but that's a nonsense really . . . we can't get them to pay much attention to other minorities too.

This, then, conveys an endemic and enduring conflict between, on the one hand, architects' concern with the apparel of the built environment, the decorative and the ornamental as signifiers, and, on the other, the concerns of those who wish to democratize the design process, to integrate the forms and uses of buildings in ways which are driven by the designers' empathy towards the users of the built environment.

In particular, the ideological nature of the aesthetic and the technical, of the architect as somehow a neutral arbiter, able and willing to provide for all, has been exposed by a range of writers who indicate how the institutional nature of the profession is dominated by a strand of conservatism which seeks to perpetuate ableist, masculine, values (Imrie, 1996a; Laws, 1994a, 1994b; Rose, 1990). As Matrix (1984) and others have argued, built spaces in the postwar period have emphasized mobility over accessibility and have placed a premium on, for example, individuals owning a car (see Women in Geography Study Group, 1988). Indeed, designers tended to generate and perpetuate exclusive, segregated spaces, primarily because of a stereotypical conception of people as somehow being similar in their capacities to both get access to and move around the built environment. Yet, clearly, this is not the case and the myth of the 'normal person', of a white male, has been a powerful dimension of the design process, yet one which has had, and continues to have, clear racist, sexist, and ableist connotations. This, then, is far from designing for the subjective being, for human diversity, in the way in which authors like Davies and Lifchez (1987) call for.

It is clear that such exclusions were, and still are, enshrined and maintained by virtue of the institutionalized nature of the architectural profession (see Knox, 1987). A significant part of this relates to its wider governing bodies, especially the architectural schools and other regulatory bodies which have the primary responsibility for overseeing professional practices and conduct. Indeed, a range of literature indicates how the governing, corporate bodies, like the Royal Institute of British Architects (RIBA), have been complicit in reinforcing the elitist structures of architecture, and, as Lifchez and Winslow (1979) have argued, while the proportion of the population with disabilities grows, the architectural profession has been slow in taking account of the environmental implications of an ageing and/or increasingly disabled population while few practitioners, less than four per cent in the USA, even less in the UK, have a disability (Blytheway and Johnson, 1990). Indeed, even where perspectives on disability are taught in architectural schools, they are still treated as an after-thought, an add-on, and/or a special-interest subject, or what Davies and Lifchez (1987) have referred to as being underpinned by a 'system of indifference'.

Thus, for instance, the RIBA's 1993 curriculum briefly mentions disability in Parts 1 and 2 of the Design Studies examinations. Students are told that examiners will be looking at the ways in which they have 'interpreted and worked within the brief' which includes taking account of 'disabled movement within the building' (p. 23). Indeed, in interview, a respondent from the RIBA's education department noted that 'part of the problem is that so few disabled people do our courses and so present no direct challenge to the system . . . it's easy just to ignore their concerns'. Such exclusions are also reinforced by the fragmentary nature of the architectural profession, a situation which is epitomized by the inability of the RIBA to exercise corporate regulation and/or controls over the practices of its members and, as an RIBA spokesperson commented in interview, 'our practitioners are sole practitioners, it's all private and self regulating . . . the only regulatory body is the Professional Standards Committee but they're weak and only interested in financial irregularities' (also see Wates and Knevitt, 1987). Moreover, while the RIBA stipulate that their members are required to take Continuing Professional Development courses, equivalent to thirty-five hours per year, the respondent admitted that they 'don't need to declare to us that they've done it and we don't actually monitor the system'.

The relative absence of corporate controls over architects, then, relates to some extent to the privatized nature of architectural practices. Yet, while disablist design can, in part, be understood as being perpetuated by the fragmentary nature of this system, a crucial aspect of our understanding also relates to the hierarchical and elitist nature of privatized clientist patronage that architects are locked into and, crucially, to related systems of economic power (Dickens, 1980; King, 1984; Knesl, 1984; also see Wright, 1991). As Crawford (1992) and others have documented, the peculiarities of the rise of the architectural profession left architects more or less wholly dependent upon a small group of clients who could afford to support them and their ambitions (Ghirardo, 1991; Jencks, 1987). Thus, while some architects gained status and economic remuneration, primarily by being sponsored by business and corporate capital, their autonomy was (and still is) heavily circumscribed by their clients (for an extended discussion of this see Crawford, 1992). As Crawford notes, 'architecture, a luxury rather than an indispensable service, remained within a pre-modern model of elite patronage, its provision of services primarily dictated by economic power' (p. 31).

Indeed, as Crawford recounts, the dependence of architects on the wider corporate economy was a determinant of their loss of technical and economic control over building projects and, by the early 1950s, the combination of systems building, new technologies and the rise of the global economy was beginning to undercut both their status and levels of autonomy (King, 1984; Knox, 1987; Sennett, 1990; Wolfe, 1981). As Pawley (1983) and others have commented, the estrangement of architects from the wider building processes was paralleled by the emergence of a division of architectural labour which drew increasing numbers of architects into managerial and bureaucratic roles, while reducing the amount of new building being commissioned through architectural practices (Knox, 1987). In this sense, the extent to which architects, and other building designers, were exercising control over the built form was increasingly being challenged by a range of structural factors and, as Crawford (1992) has noted, it is the materialities of the land market, or 'the actualities of the building industry, and the limits set by the clients paying the bills' (p. 38), which have become the dominant element in restraining the autonomy of architects and/or designers.

Thus, the emergence of corporate economics was, as Jencks (1987) points out, crucial in perpetuating the move towards technological standardization and scale economies in building design, while seeking to realize cost savings by utilizing cost efficient building methods which made few concessions to the range of users who did not conform to the conception of the able-bodied client. The evolving architecture was increasingly cast in the form of a globalizing economic imperative, the emergent form exuding a functionalism indicative of the rise of corporate power. In particular, the new (corporate) spaces were simultaneously public and private or, in Jencks' (1987) terms, the new functionality was premised on drawing in those with economic power, excluding those without. Ornamentation persisted in a minimalist way, yet sufficient to reinforce the classical conceptions of built form, of the raised steps, split levels – in effect, modernism had reached its apogee.

THE MODERN IDEAL – THE DISABLING SPACES

While the contemporary western city is characterized by a pastiche, a constellation of diverse styles and forms, the dominant ideas and practices which have done most to shape it are clearly linked to modern aesthetic philosophies and socio-political institutions (Giddens, 1991; Jencks, 1987; Knox, 1987; O'Neil, 1995; Wolfe, 1981; Wright, 1991). The architectural and processual styles of modernization were initially espoused in the earlier part of the twentieth century and, as Wright (1991) notes, they represented a self-consciously linked definition in which, as Wolfe (1981) argues, there was a historical inevitability to the forms and styles that should be used. Indeed, modernism, as the aesthetic expression of modernization, had its intellectual roots in the modern avant-garde which characterized modernization as a progressive and inevitable force, one borne out of a wider social movement resting on the 'axiom of human domination of nature', of the supremacy of naturalistic scientific values, of human rationality and will (O'Neil, 1995, p. 156). In particular, modernism was presented as the antithesis to bourgeois culture in seeking to generate a new collectivism based on the dismantling of what Knox (1987) has termed the 'capitalist prescriptions of power and extravagance in urban design' (p. 353).

Yet, while modernism was premised, in part, on the idea of liberation, of universal freedom, its claim to transcend politics and power, to overturn the bourgeoisie, was never evident in the developments of postwar urbanism, characterized by what McGlynn and Murrain (1994) have termed the advent of the segregated and mono-functional forms, an aesthetic closely aligned to the rise of the corporate economy. Indeed, the engineering aesthetics of the modern movement were, as Weisman (1992) notes, built upon an abstract, intellectual purity of rational, geometric forms, and a mass produced industrial technology. Any sense in which it could relate to differences in body, human behaviour, or access requirements were all but lost in a style that many have referred to as 'non contextual' architecture, premised on forms which seemed to deny human subjectivity and the differences in bodily experiences and forms. In 1929, for instance, the English architect, Eileen Gray, characterized such non contextuality in the following terms:

> this intellectual coldness which we have arrived at and which interprets only too well the hard laws of modern machinery can only be a temporary phenomenon . . . I want to develop these formulas and push them to the point at which they are in contact with life . . . The avant-garde is intoxicated by the machine aesthetic . . . But the machine aesthetic is not everything . . . Their intense intellectualism wants to suppress everything which is marvellous in life . . . as their concern with a misunderstood hygiene makes hygiene unbearable. Their desire for rigid precision makes them neglect the beauty of all these forms: discs, cylinders, lines which undulate or zigzag, elliptical lines which are like straight lines in movement. Their architecture is without soul (quoted in Nevins, 1981, p. 71).

In this sense, modernism was founded upon the idea of the minimalist building and/or design bereft of (bourgeois) ornamentation or, as Wolfe (1981) comments, buildings were to express function and structure and nothing else. In particular, the movement which grew up around such ideals, including the Bauhaus school, Les Congres Internationaux d'Architecture, Archigram, and the Ekistics school, asserted the importance of science and technology in the production of the built form, of the need to build inexpensively, to provide for all in the community. Its clarion call was originally espoused by the American architect, Louis Sullivan, that form should follow function, a maxim which many interpreted as the search for universal laws of human habitation and behaviour, of the possibilities of producing 'pure' design, singular styles and forms, which were grafted from the essence of the human being. In this sense, functionality was expressed as a means of maximizing building utility, premised upon the idea that human behaviour was wholly predictable and knowable, that human beings conformed to a type, to particular patterns of (able-bodied) normality both in bodily and mental terms. Thus, human beings were, in this conception, reducible to a specific essence, an essence which, as we shall see, was the embodiment of ableist thinking.

In particular, the ableist nature of modernist ideas is revealed, in part, by its conception of functionality whereby there was a departure from seeking an individual or specific solution for what Sullivan (1947) termed 'a true normal type'. The search for such normality was evident in the thinking of one of the leading exponents of modernism, Le Corbusier, who believed that the propagation of universal properties in form giving was an essential underpinning of the architect's mission, or, as he commented, 'all men (*sic*) have the same organism, the same functions . . . the same needs' (1927, p. 27). This search for normality, an inner essence, in people provided the context from which a distinctively modern movement, or interpretation of form follows function, evolved. As Le Corbusier noted:

> the establishment of a standard involved evoking every practical and reasonable possibility and extracting from them a recognized type conformable to all functions with a maximum output and a minimum use of means and workmanship and material, words, forms, colours, sounds . . . (p. 27).

The discovery of this 'standard', then, was at the root of the modern preoccupation with function, and, as Le Corbusier (1927) argued, the bare essentials of architecture are provided by aesthetic forms which he defined as being 'determined by the dimensions of man (*sic*) and the space he occupies' (quoted in Gardiner, 1974, p. 79; also see King, 1984; Sullivan, 1947). For Le Corbusier, architecture could only be defined in and through the symbiosis between people and nature, and, as he commented, 'man must be rediscovered' (quoted in Gardiner, 1974, p. 79).

Yet this rediscovery was wholly based on a particular, ableist, gender-specific, conception of the person, an idealized man who was presented as the embodiment of normality. This embodiment of normality was expressed in a diagram conceived by Le Corbusier in 1925 called the Modular, a device which utilized the proportions of the (able) body to enable the architect to create the built spaces, or, as Le Corbusier argued, 'one needs to tie buildings back to the scale of the human being'. Yet, as Figure 4.1 indicates, the Modular presents an image of an upright person, muscular, taut, obviously strong, male, and displaying no outward sign of either a physical and/or mental disability. It is the person for whom functionality in building design and form was being defined, a person who gained widespread acceptance in most elements of the modern movement and beyond. Indeed, the parallels between Le Corbusier's Modular diagram and that presented by Dreyfuss (1955) are striking, although

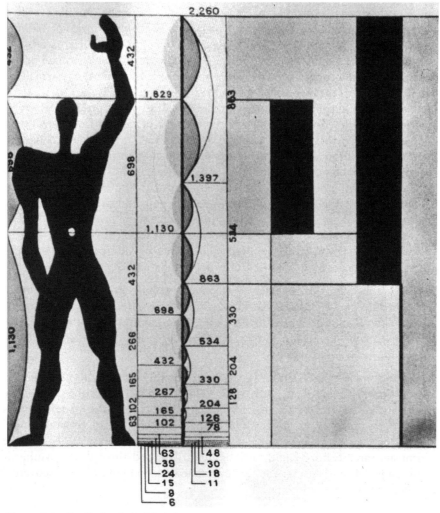

Figure 4.1   Le Corbusier's Modular

Dreyfuss does concede that women exist and have different physical attributes to men (see Figure 4.2). Yet, the overall effect complements that of Le Corbusier, the denial of bodily diversity and differences and the projection of normality as one of able-bodiment.

Such conceptions can also be extended to incorporate the possibilities that the denial of bodily differences was also premised on the idea of asexuality as the moral, or ethical, standard bearer of the emergent aesthetics of modernism. Indeed, as Batchelor (1994) comments:

> authentic modernism, predicated on the extent to which it excluded, was probably white, male, and uneasy with sexuality. Many modernists addressed the emotive, sensuous aspects of experience and the possibilities that these opened up in terms of modern architecture and design. But only certain forms were licensed. Others were regularly rejected as anti-rational, barbaric and representing a retreat to the primitive (p. 115).

Indeed, as O'Neil (1995) has noted, one of the ironies of the modernist project was the way in which its rationalism abstracted from the socio-political contexts of its practices, failing to communicate, or interact, with those who were the (often unsuspecting) recipients of the resultant built forms. As Knox (1987) has commented, how then could modernism ever hope to know of the subjective experiences of the users of the built environment when its philosophies more or less discounted the realm of the experiential, personalized, experience? As Wolfe (1981) and Knox (1987) have described,

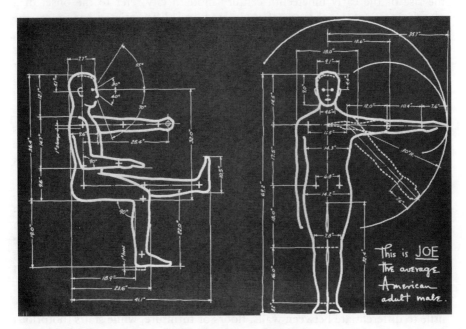

Figure 4.2 Henri Dreyfuss and his conception of the body

the leading exponents of modernism – Gropius, Le Corbusier, Mies van der Rohe, even its critic, Robert Venturi – asserted the 'special insights' of the architect, of their privileged access to knowledge. For instance, as Mies van der Rohe once replied, when asked if he ever submitted alternative schemes to clients:

> Only one. Always. And the best one that we can give. That is where you can fight for what you believe in. He doesn't always have to choose. How can he choose? He hasn't the capacity to choose (quoted in Prak, 1984, p. 95).

In this sense, modernism was underpinned by a 'theory of technocracy, government by experts, rather than democracy, government by people' (Steinberger, 1985, p. 39). Consequently, the (claimed) intellectual purity of modernism descended into forms of what Knox (1987) has termed 'vain arrogance' or the perpetuation of an elitism whereby 'clients, other professionals, and users were systematically excluded and often patronized' (p. 369).

Not surprisingly, then, the abstractions of modernism, coupled with an elitist philosophy, were, in part, at the root of the inability of people to influence the processes of production of the built environment. The emphasis on sameness, on uniformity, was problematical for its failure to differentiate between users and to recognize that places and spaces need to be multifunctional to cope with human diversity. In particular, commentators on the rise of the early nineteenth century modern city refer to the onset of a placelessness, of an absence of variegated and differentiated spaces, of the dearth of place markers and/or signifiers (Giddens, 1991; Harvey, 1990; Savage and Warde, 1993; Wirth, 1936). In addition, a placelessness was seemingly underpinned by what Sennett (1990) has argued to be the modernist pre-occupation with notions of functionality and wholeness which, as he suggests, generated conflicts between buildings and people, not the least of which is that 'the value of a building as a form is often contrary with its value in use' (p. 98). This observation prompted Sennett to refer to modernism as 'bequeathing the anti-social building' precisely because he saw the problems being generated by virtue of the irreducibility of human diversity to the types of environments which were being created.

Indeed, in terms of the legacies of the movement, of style and form, modernism has been characterized by designer ableism in a number of respects. As Moore and Bloomer (1977) have noted, for instance, in modern architecture the multiple changes of level 'have often been used to delineate and enliven space' yet in ways which elevate the aesthetic above the pragmatics of use (p. 4). Thus, the interplay between levels, connected by steps, is integral to a design which seeks to display divisible, yet interconnected, functional, spaces. Moreover, the minimalism influencing much modernist design does little to differentiate between walls, floors or furniture, while stairs (notorious barriers to mobility and access) have often been given symbolic roles. Indeed, as Salman (1994) has argued, the main effect of the Bauhaus movement, one of the lynch-pins of modernism, was to reduce space, to automate, and to utilize standard, off the shelf materials, primarily in an attempt to persuade people into using a certain kind of pre-determined design. As Figure 4.3 shows, everything in the room is seemingly fixed and unchanging, displaying the possibilities of a mass produced (and consumed) architecture. This, then, is an example of design that matches visible, so-called able-bodied, people: the small kitchen, the knob hardware, and rooms which are relatively easy to fabricate yet more or less impossible to get into.

Figure 4.3   An illustration of modern kitchen design

While the interiors of dwellings, and their immediate exteriors, often served to exclude people with disabilities, the dehumanizing and disabling elements of modernist ideas are also evident in a range of other contexts. One of the best documented, although not in relation to disablist design, is the British new town (Hall, 1984; also see chapter 6). Their theoretical underpinnings, in part, borrowed ideas from Le Corbusier's publication, *La Ville Radieuse*, which espoused the (functional) merits of the spatial separation of different land uses, the formation of the zoned city. For Le Corbusier and others the separation of the home from the workplace, and of industrial uses from parkland, was indicative of the 'good city', yet such ideas were premised upon a spatial separatism which placed heightened emphasis on people's mobility as a means for them to gain access to the component parts of the emergent city environments. The resultant built forms were not neutral in their social effects and, as a range of researchers has documented, the 'good cities' were actively discriminatory because not everyone could be equally mobile, while the provision of intra-urban public transport, which was one possibility of enabling people to overcome the 'frictions' of distance, could never be guaranteed (Matrix, 1984; Wajcman, 1991).

Yet it would be unfair to characterize Le Corbusier, and, indeed, some other modernists, as entirely disablist in as much that there was a recognition of, if not direct designing for, what was regarded as the 'subversive' potential of people. Thus, Le Corbusier's design for workers' houses at Pessac in 1925 was characterized by the assembly of windows, staircases and other parts of the buildings in non standardized forms with flexible open spaces left for the individual occupiers to impose their own design values. As Le Corbusier acknowledged, 'this, of course, destroys the visual consistency of the architecture . . . you know, it's life that's always right and the architect who's wrong' (quoted in Jencks, 1987, p. 74). In interpreting Le Corbusier's contradictory feelings about Pessac, Jencks (1987) has argued that:

> Starting with the idea of resolving two incompatibilities like the individual and the group, it was not surprising that Le Corbusier could end up, as at Pessac, by admiring the way personalization was destroying his own architecture. All the arguments for a geometrical civilization . . . were countered by the barbaric actions of the inhabitants at Pessac, and yet, according to the supreme dialectician, these barbarians were still right (p. 74–75).

Yet, Le Corbusier's notion of 'rightness' was still circumscribed by the idea that people, by transforming their living spaces, were somehow, in his terms, 'subverting and undermining' the ideals and purity of the architect, although, as Knox (1987) has noted, nowhere to the same extent as the corporate client who saw the possibilities for combining the modernist ethic of minimalism, of standardization, with cost savings in the production of the built environment.

Indeed, as Knox (1987) has observed, the rise of the corporate economy helped to foster modernism and, in some senses, coopted it as a self serving means for legitimizing efficiency and cost savings in building projects. In this sense, the interpenetration of modernism by wider corporate, economic, values introduced a form of stasis into the built environment. In particular, the rise of Fordist production methods, the move towards standardized technology and the emergence of a task-dedicated division of labour were all part of the wider ethos that science and technology could facilitate corporate control, that people were reducible to specific types. For the modern corporation, the idea that people were 'all of a type', that they too could be standardized like a piece of technology, was, in part, incorporated into the lay-out and design of the emergent work-places, and as Sennett has observed, modern buildings are less flexible than the 'rows, crescents, and blocks of the past', while the specific layout of the modern office environment is task-dedicated and more or less impossible to change towards alternative types of uses. In this sense, the sponsors of modern designers and architects, the corporations, were implicated in placing a demand for an ableist environment to cater for an able-bodied workforce, an exclusion upon an exclusion.

Ultimately, the underlying socio-political and institutional relations of modernism have served to exclude and deny minority people's access to the centres of power where the crucial decisions are made about the production of the built environment (Giddens, 1991; King, 1984). As Walker (1994) has powerfully observed, 'it's distasteful . . . we live under a system of architectural apartheid. I'm like anyone else. I want a choice. And respect. I don't want to have to ask to get in and out of buildings and buses . . . what would it be like if black people, pram pushers, or homosexuals couldn't use public buildings.' Yet, while modernism has failed, the extent to which the promises held up by an apparently new epoch of social and political organization, that of postmodernism, in overcoming socio-spatial inequalities like disablism, is itself questionable, a theme I now turn to.

BEYOND MODERNISM – THE PROSPECTS OF AN ABLEIST POSTMODERNISM

A range of writers has argued that there is a crisis in modern society, characterized by the breakdown of its socio-cultural traditions, the emergence of social tensions underpinned by the fragmentation of labour markets, and the development of new socio-spatial inequalities (Giddens, 1991; Jenkins and Gray, 1991; Lash and Urry, 1987; Walker and Sayer, 1993). Giddens (1991) highlights the paradoxes associated with what he terms high modernity, a period of change which does not 'consistently conform to human expectations or to human control' (p. 38). In particular, events like Chernobyl in 1985, the explosion of the Challenger spacecraft in 1986 and the ravages of pollution in the aftermath of the 1991 Gulf war have all served to highlight the vulnerability of western beliefs in the ability of people-devised technocratic systems, founded on a seemingly infallible (scientific) knowledge base, to deliver wholly controllable and predictable socio-environmental contexts.

Such circumstances, since the late 1970s, have also engulfed architecture to produce what Jencks (1987) has termed a 'crisis of meaning', characterized by the loss of confidence in the certainties promised by modernism, of stability, order, and stasis (see Sennett, 1990). Indeed, as Jencks (1987) and Knox (1987) have documented, even before this period, disquiet was being expressed about modernism in the form of a return to neo-vernacular themes, while the influential criticisms of Venturi (1967) were based on the call for more referencing to hybridity, to reflect complexity, contradiction and difference in the built form. Thus, by the early 1980s, a postmodern architecture seemed to be in the ascendancy 'characterized by an engagement with subjectivity, and by attempts to restore meaning, rootedness, human proportions, and decoration' (Savage and Warde, 1993, p. 139). As Savage and Warde noted, the emergent postmodern condition was premised on notions of specificity, of localized scale and context, a celebration of local vitality, of the need to restore the human dimension.

Potentially, then, as Moore and Bloomer (1977) have argued, the emergent postmodern spaces contained the basis for overturning the dehumanizing effects of modern architecture, of environments which are, as they allege, characterized by paternalistic signs and structures. Indeed, the underlying philosophy of postmodernism – a rejection of totalizing theories and discourses, a recognition of the fragmented and discontinuous nature of the world – seemed to provide for the possibilities of a political engagement with the dispossessed and disenfranchised. As Knesl (1984) claims, postmodern architecture represents 'an answer to the distraction, ennui, hostility, and powerlessness of contemporary urban society', so suggesting that oppressed groups might be able to contest, and claim for themselves, the emergent built forms (Knesl, 1984, p. 16, quoted in Knox, 1987, p. 361). Others have noted how the postmodern is fluid, not fixed, and open to all who wish to influence it, although they are less illuminating on how politically marginalized peoples can grasp the requisite power in a period of (alleged) flux and uncertain and shifting values (Dear, 1986, 1995; Giddens, 1991).

Indeed, there is little to suggest that the material power structures of architectural production are necessarily breaking down, nor that the emergent architectural forms are any less hostile to people with disabilities than those which are (supposedly) characterized by modernist ideals. Jameson (1991) describes the confusion of the new spaces of postmodernism, or 'the profusion of curiously unmarked entrances to the buildings, all lateral and rather backdoor affairs' (p. 45). In commenting on the Bonaventure hotel in downtown Los Angeles, Jameson remarks on the disorientating and disorganizing spaces, 'the reflective exterior which achieves a peculiar and placeless disorientation of the Bonaventure from its neighbourhood' which transcends the 'capacities of the human body to locate itself, to organize its immediate surroundings perceptually, and cognitively to map its position in a mappable external world' (p. 45). For Jameson , the experience of postmodern space is disorientating and disabling:

I am proposing the notion that we are here in the presence of something like a mutation in built space itself. My implication is that we ourselves, the human subject who happens into this new space, have not kept pace with that evolution; there has been a mutation in the object unaccompanied as yet by any equivalent mutation in the subject. We do not yet possess the perceptual equipment to match this new hyperspace, as

I will call it, in part because our perceptual habits were formed in that older kind of space I have called the space of high modernism (p. 39).

In this sense, it seems to be the continuities, not the discontinuities, between the modern and the postmodern city which define the day-to-day lived experiences. As Harvey (1990) has argued, much of what we term postmodern is redolent of shifts in style, forms of, and contexts for, consumption. In particular, Savage and Warde (1993), echoing Harvey (1990) and Jameson (1991), refer to the emergent spaces of postmodernism as 'placeless realms', producing a built environment, like the shopping malls, that once inside 'one could be almost anywhere in the world; links to other parts of the urban fabric seem tangential and haphazard' (p. 140). Such spaces, then, seem divorced from their local contexts producing disorientation, confusion, and leading Jameson to conclude that the environment is a messiness 'in which things and people no longer find their place' (p. 117–118). Indeed, the new spaces increasingly reflect the affluence of the middle classes and their enhanced levels of mobility and it is hardly coincidental that the last ten years have been characterized by the intensified hollowing out of the city and an increasing dispersal of functions.

Such tendencies, then, highlight the contradictory nature of the claims made by postmodern thinkers about the emergent urban form and, as Savage and Warde (1993) claim, there appears to be more that binds the postmodern to the modern than separates them. As Savage and Warde (1993) note, much contemporary architecture is a mutated form of modernism, while the popularization of the warehouse and out-of-town retail unit exhibits forms of functionality (and uniformity) of the type commonly associated with the period of modernity. There is also little evidence to suggest that the institutional practices of architecture and building design are any more open to the views and ideas of lay communities, while there is no evidence to show that people with disabilities have broken down the ableist structures of the design professions. Indeed, as O'Neil (1995) notes, the era of the so-called postmodern has been characterized by the decline of the public realm, the privatization of public spaces, the dismantling of welfare states, and the emergence of non elected local government, all of which seems anathema to forms of political emancipation for people with disabilities.

This, then, places a different interpretation on the nature of the emergent postmodern landscapes and their relationship with oppressed and politically marginalized minorities. Far from the social relations of post modernism opening up the means for such groups to influence the powerful, the evidence suggests, if anything, the reverse. As Giddens (1991) argues, we seem to be in a context of high modernity where a heightening of social control by elites and the powerful is a key signifier of our existence and where the politics of postmodernism, the espousal of differentiation and fragmentation, provide a rationale for states to challenge universal welfare spending and provision (O'Neil, 1995). So while, in theory, postmodern thinking provides for the possibilities of a liberation of the individual, of a heightened sensitivity to differentiated values, the reality tends to be one of relativist propositions, the acknowledgement, indeed, the insistence, that there is no such thing as a dominant core value. In this sense, the assertion that society and its architecture are disablist would, necessarily, be rejected.

In particular, Frampton (1992) develops the argument that what we term the post-modern reaction to the modernist movement really represents the 'aestheticization of late modernism as a compensatory strategy' (p. 23). Frampton refers to the pastiche and the use of the neo-vernacular as a form of corporate packaging 'to provide nothing more than a set of seductive images with which to sell both the building and its product' (p. 23). For him, there is little to suggest that the new built forms are in any way sensitized to, or empathetic of, the access requirements of people with disabilities, or any other person for that matter. Indeed, far from the emergent post modern landscapes representing a new heightened humanity, and/or a means of liberating the disenfranchized, if anything they are more de-humanizing for holding out promises that can never be delivered. Indeed, for Callinicos (1990) the discourses of postmodernism are constitutive of class relations and are the outcome of a socially mobile intelligentsia seeking 'to articulate its political disillusionment and its aspirations to a consumption-oriented life-style' (p. 115). For the economically marginalized, this world holds out little or nothing.

In this sense, as Knox (1987) notes, the emergent architecture reflects, in the main, a postmodernism of restructuring, the built environment as a mechanism of recapitalization and recommodification, and the assertion of market values. For Cooke (1988) postmodernity means coercion 'whether that of the enforced redivision of labour which deindustrialization, new technology and casualization bring, and that of the market place with its segregating and enclosing effect on social groups' (p. 490). Thus Cooke (1988), quoting Davis (1985), describes the evolving postmodern landscape:

> the post-modernist towers are, in such a context, fortresses protecting the new rich from the new poor whom they nevertheless need, but at arms' length. Thus the hyperspaces to which access by pedestrians is all but impossible become comprehensible. And the metaphor is completed in the later styles to be found in construction in Manhattan where new towers are coned and crenellated, and in one case surrounded by a moat and provided with drawbridges (p. 110).

This, then, is suggestive of oppressive environments, of exclusive and hostile spaces, of a postmodernism which is predicated on, and seemingly underpinned by, the reaffirmation of the coercive values of society (of racism, sexism, disablism). Indeed, even if postmodern spaces are physically accessible (which is questionable), Cooke (1988) and others doubt whether the social spaces of (corporate) postmodernism have transcended the hierarchies, or the closures, of the technical and social division of labour, hierarchies which say 'this space is ours'. In this sense, the liberating potential of postmodern epistemologies, and their related practices, are questionable.

TOWARDS UNIVERSAL DESIGN AND AN EMANCIPATORY ARCHITECTURE?

While modernist ideals are alienating and fundamentally ableist, alternative ways of thinking and practising exist and there is a range of socio-architectural practices based on a contrasting set of ideas and philosophies about the interactions between humans and the built environment. Perhaps the most widely espoused is that of universal design, an approach to the construction of the built environment premised on, as Weisman (1992) calls it, a 'flexible architecture' or one based on structures which are 'demountable, reasonable, multifunctional, and changeable over time' (p. 32). As Weisman notes, the (modernist) construction of the built environment conceives of spaces as somehow fixed and unchanging, while buildings have tended to be (and still are) dedicated to single functions, creating a form of stasis or unchanging places. However, people and places are fluid, transformative and multi-dimensional, yet much architecture seeks to fossilize specific forms of social relations while denying, even resisting, the dynamic nature of society.

The reaction against such conceptions represents one of the real strengths and contributions of postmodern thinking by emphasizing the vitality and importance of other cultures and values over and beyond hegemonic discourses, and the need to generate political spaces for their articulation. In part, such conceptions underpin universal design or what Salman (1994) defines as a viewpoint which states that environments should be sensitized to all users, that there is no such thing as stasis in the built form, and that flexible building designs should be utilized to permit people to transform the fabric of the places and spaces that they interact with (in). Others see such principles as being trans-generational while incorporating choice and alternatives into the built fabric. Indeed, as Davies and Lifchez (1987) note, accessibility is much more than admittance to a building or a matter of logistics but is also a quality of socio-psychological experiences which modernist ideas did little to acknowledge. They comment:

> how one feels about a place, how one interprets it, or even whether one can adequately interpret it – these are all less quantifiable, but crucially important, aspects of accessibility. A place that supports people's activities and desires permits them to be and do what they want, and causes them a minimum of pain, frustration, and embarrassment is more accessible than a place that confuses, harasses, or intimidates people. Many ostensibly accessible sites differ substantially in the quality of experience they offer (p. 40).

Principles of design which reflect this wider conception of accessibility while denying the stasis of the built form are evident in a number of places, and especially in the Netherlands, Germany, and France where a range of housing schemes have been developed with the objective of being adaptable to social change (see Daunt, 1991). Weisman (1992), for instance, cites the example of Stichting Architectin Research (SAR), a Dutch-based approach to housing design based on the idea that a dwelling represents much more than a physical entity but is a 'human act'. As Weisman recounts, the future tenants of SAR schemes are involved in designing their own living spaces and, in one instance, a family living in one of the housing schemes was allowed to lower the windowsills in their living area to provide their father with a view from his wheelchair. Likewise, one example of accessibility wholly incorporated into a new building is the APL building situated in downtown Oakland in California. As Figure 4.4 indicates, accessibility for wheelchair users has been designed-in to the front entrance yet, as the person responsible for access compliance commented:

Figure 4.4   The APL Building represents a design 'success' for people with disabilities in that it is fully accessible for most people with physical and/or mental impairments. The pavement approaching the building has a deep pink colour to permit visually impaired people to differentiate between the line of pavement going past the building and that directing individuals into the building. Entrances, as the photograph shows, are wide with shallow gradients.

at the time of the original plans, the architects were insisting that the look of the building would be compromised by the front entrance access ways we were insisting on. They wanted it all in a side entrance and they see this as good enough but our attitude is that people with disabilities have the same rights as everyone else to go through the front door!' (Gertner, 1994).

The conflict, in this instance, was the issue of the aesthetic versus the humane, of facilitating access for all types of persons, or prioritizing particular design aesthetics which, so APL claimed, were a necessity to the 'corporate image'. Yet, as Gertner (1994) argued, 'to see the two issues as separate, or in the way APL was presenting it, was crazy . . . they've still got their image and we've got some access'. While all built forms are to a certain extent fixed, the APL building is interesting as an example of a design with removable internal walls and floors, while it incorporates induction loops and other technologies to facilitate movement around the building for different types of people with disabilities. However, its production occurred within the hierarchical relations of a corporate sponsor and a commissioned architectural practice and, as a range of researchers has noted, how far is it possible to produce sensitized design in a context where the social relations of building production are largely removed from democratic control and popular involvement (Davies and Lifchez, 1987; Knesl, 1984)? As Davies and Lifchez (1987) have questioned, 'how ethical is it to practice architecture, to be a professional licensed to design buildings, without having first developed an intellectual and emotional understanding of people' (p. 35)? In this sense, Davies and Lifchez suggest that architects need to confront the social psychological context of design, how it feels for the users and to acknowledge that there are no simple technical (design) solutions.

This then suggests that a people-driven, or derived, community architecture is important, of the type illustrated in Figures 4.5 and 4.6, a playground in San Francisco. The playground is located in the poor Tenderloin district in the downtown and was designed by a team of local people with all forms of physical impairments. People with disabilities living locally had campaigned for the playground for a number of years, insisting that they had the capacity to design it. As the access officer for San Francisco (1994) said, *they kept berating me to get it going and were insistent that they had to design it for their children'*. Children were widely consulted about its design, while the access officer for the city (pictured in the photographs) spearheaded the initiative by allocating a proportion of his department's budget towards the implementation of the scheme. The fixtures in the playground are interchangeable, while colouring and tactile surfaces provide 'guidelines' for people with visual impairments. In part, it illustrates one of the ideals of universal design, empowering local communities and creating the structural contexts within which a form of emancipatory, liberating and self determined practice can occur. Yet, as the access officer for San Francisco argued, 'it was

Figures 4.5 and 4.6   Richard Skaff, the access officer for San Francisco, demonstrates the ease of access to the playground purpose built for children with disabilities. Sensitively placed fixtures permit some wheelchair users to hoist themselves up onto swings and slides while the whole surface is constructed out of soft materials to minimize injuries. As Richard Skaff commented, 'the place is very well used by all sorts of people, not just children with disabilities . . . and you get lots of disabled kids here mixing with the rest'.

still funded by us, we brought in some architects from outside too, we had to keep some control over the way the budget was being spent'.

Indeed, while, in this example, people were being empowered at the level of formative ideas, forms of estrangement were still apparent and practices were not being devolved and/or transferred. As O'Neil (1995) has argued, part of the problem of estrangement more broadly is related to the professionalization of a whole range of social activities, so maximizing the bureaucratic ethos and undermining the civic competence that many feel is a pre-requisite of democracy. In this sense, one way of returning knowledge and competence to the wider community is, as O'Neil argues, to institutionalize the 'transferability and thereby accountability of expert knowledge in order to raise the level of the well informed citizen or the need to create a pedagogy that will subordinate expert knowledge to the needs of political democracy' (p. 170). This, then, is one example of what Oliver (1992) terms reciprocity, or a situation whereby the role of the architect is to be an enabler and educator rather than preacher or provider, with the resources of the design industry being placed at the disposal of local communities (see Wates and Knevitt, 1987). Yet this is only one dimension, and a key absence in design thinking relates to the embodiment of meaning in the built form, or that which is literally derived from our bodily experiences. As Moore and Bloomer (1977) have argued, in relation to the visually impaired:

> the historic overemphasis on seeing as the primary sensual activity in architecture necessarily leads us away from our bodies. This results in an architectural model which is not only experientially imbalanced but in danger of being restrictive and exclusive . . . especially when we consider that all sensory activity is accompanied by a bodily reaction (p. 6).

Thus, to empower people with disabilities in the design process is a multi-faceted, multi-dimensional process which, as a minimum, requires an engagement at the level of values and ideology, as well as the material base of building processes. Indeed, as Knesl (1984) notes, one needs to radically rethink the 'relationships between both the architect and the client and between the architect and the process of designing a building' (p. 4). For Crawford (1992), for instance, the crucial issue is how to re-connect architecture to socio-economic materialities, to reformulate its theories and practices in order to transcend what many regard as a step back towards the identity of architecture purely with aesthetic concerns. As a first step, Crawford suggests that the architectural profession needs to establish new relations with the technical and economic practices of building, to analyse the material conditions and abandon their reliance on technical capabilities as the mechanism for advancing them. In particular, Crawford calls for architects to envision a new set of ideal clients, especially those who are estranged and displaced from what he terms the architectural market place.

However, part of the problem with Crawford's formulation is the way in which it maintains the distinctions between the architect as 'expert and the client as somehow 'unknowledgable', without the intellect and/or means to engage with (in) architectural discourses. In particular, Crawford's conception of the 'architect's dilemma' is reducible to the recurrent issue of self identity and legit-imation and has less to do with recognizing and analysing the nature of the very social issues which Crawford feels that architects should be concerned about. Indeed, the social relations of architectural production, suggested by Crawford, do little to displace architects from their privileged position. This criticism can also be extended to the intersection between postmodern epistemologies and those discourses which, while espousing the possibilities for human emancipa-tion, fall far short of specifying the material political conditions within, and through, which liberating states might ensue. Indeed, the problem of, and for, universal or other conceptions of design are their idealist bases, propagated by an intelligentsia whose social and intellectual roots do not necessarily corre-spond with the wider communities that they problematically claim to build for. For the foreseeable future, the hierarchical social relations of architectural production are unlikely to be transformed.

CONCLUSIONS

As Davies and Lifchez (1987) have argued, access should not be viewed as a constraint on architectural design but should be conceived of as a 'major perceptual orientation to humanity' (p. 49). In this sense, a range of authors note that design professionals increasingly need to reject the idea that there are technical solutions to socio-political problems, that there needs to be some kind of deconstruction of the ideological constructs that underpin the aesthetic ideals of design. Indeed, many writers concur that there can never be a socially-sensitive or just architecture given the present structural underpinnings of architectural practices (Knesl, 1984; Knox, 1987). As Crawford (1992) concludes, 'the restricted practices and discourse of the profession have reduced the scope of architecture to two equally unpromising polarities: compromised practice or esoteric philosophies of inaction' (p. 41). Yet, as Crawford suggests, such an impasse need not necessarily prevent architects from reconnecting their practices to social and economic questions, to issues, for instance, relating to the elderly, poor, people with disabilities, and the homeless. Unfortunately, one still waits for such connections to be made.

Yet, others are more optimistic in seeing the seeds of liberating environments and of the possibilities for non ableist architectural practices. Hayden (1981), for instance, considers the elements of a transformative agenda which would challenge the socially oppressive nature of much past and contemporary architecture. She calls for a 'new paradigm of the home, the neighbourhood, and the city', one which describes, as a first step, the 'physical, social, and economic design of human settlements that could support, rather than restrict, activities of people with disabilities' (p. 7). Likewise, Weisman (1992) locates the problematical aspects of access, of exclusion and segregation, in the comprehensive system of social oppression, not, as he puts it, the consequences of failed architecture or prejudiced architects. This is a crucial point because, in conceptual terms, it situates the actions and practices of agents and institutions in a wider framework of social structures, values, and ideologies and avoids a reductionism which posits that people and/or institutions are somehow, independently, to blame for the perpetuation of disablist environments. This is a theme which I wish to extend in the following chapters.

FURTHER READING

The literature on the theme of architecture, power, and the production, and reproduction, of the built environment is vast and obviously I have only been able to touch on some of the key debates. For the reader who is interested in following up some of these then the following references will be of interest to you. The book by Nick Wates and Charles Knevitt, 1987, *Community Architecture*, is a useful introduction to the possibilities of producing people-sensitive architecture and design. Sennett (1990, 1994) provides a useful overview of the interconnections between historical conceptions of space, design, and built form, while Knox (1987) provides one of the better critical overviews of the social (re) production of the built environment.

5

Creating Accessible Environments

INTRODUCTION

Since the early 1980s, western governments have increasingly acceded to the idea that inaccessible spaces and places in the built environment require some redress through public policy. This is reflected, in the UK, by the emergence of new institutional fora for developing access policies for disabled people while, in the USA, the drive towards barrier-free environments has been a staple part of institutional life since 1968. Likewise, in countries like Germany, the Netherlands, and Sweden, significant policies and programmes, aimed at creating accessible places for people with disabilities, have been an important part of their welfare states, yet most, if not all, responses, have been piecemeal, ad hoc and poorly resourced, while tending to be an add-on to social welfare policies rather than an integral and integrative part of them. Even at the supranational, European Union, level, the emergent policy frameworks have tended to emphasize socio-technical solutions towards access, as though a transformation in design, in and of itself, will provide the singular mechanism for overturning disabling environments.

That access policies and programmes should exist at all is a recognition of the hostile and oppressive nature of the built environment and, as Barnes (1991) suggests, people with disabilities have made some gains in recent times. Such gains, particularly in the UK context, are encapsulated in a range of legislation which specifies that 'reasonable provision' should be made for disabled people's access, while, since the late 1970s, systems of professional advocacy and representation have emerged as the state's response towards correcting the seeming powerlessness of disabled people and their exclusion from debates about the built environment (Oliver, 1990). In turn, this has generated what one might term an 'access industry' with access officers and committees springing up all over the UK, co-ordinated and orchestrated by national access organizations.[1] Yet the overwhelming impression is that the plethora of policies and programmes for access, particularly in the UK, are doing little more than reflecting and reproducing elements of state welfarism, the idea that what people with disabilities are receiving (yet again) is another form of government benefit (and, so some would say, a 'handout').

For Young (1990), the institutionalization of specific group demands and political claims relating to issues like access is, in part, indicative of the (re) assertion of ableist values, of conceding to the ethical principles of facilitating access (for all), yet in ways which really fail to grasp the issue that much more than reformist measures are required (also see Wolfe, 1977). Indeed, for Gooding (1994), the British state's response towards facilitating access is highly problematical because, as she argues, it tends to define 'improved access as desirable but not a social imperative' (because, as the British government maintains, it is not a right) (p. 14). Thus, the attempt to cultivate systems of participation of people with disabilities in access issues seems to be not much more than an extension of charity or a 'hand-down' from the able-bodied professionals to those who should be seemingly grateful for any form of participative democracy. Indeed, for Wolfe (1977), such 'participative' forms are indicative of a 'franchise state' with government authority being co-opted for institutionalized interest groupings who are effectively depoliticized (and pose little or no threat to the dominant socio-cultural values of ableism, racism, sexism, etc.) (also see chapter 7 where I extend this argument).

In developing such notions, this chapter considers the critical role of access policies and programmes in the UK in addressing disablism in the built environment. I develop the argument that the institutionalization of disabled access, while drawing some attention to issues of marginalization and oppression facing disabled people, is systematically failing to redress the oppressive conditions which confront disabled people in the built environment. In part, this is because access policies are targeting the symptoms of the problem and not the underlying causes. In particular, the socio-institutional structures within which access practices and policies are being conceived are reinforcing expertism and clientism and the idea that people with disabilities are somehow a group 'to be dealt with'. Moreover, the underlying legislative basis of access, as chapter 3 has intimated, tends to be weak and reactive, while maintaining the ethos and efficacy of market utility. In this sense, as the chapter will argue, access policies are potentially implicated in the (re) production of the multiple dimensions of social oppression and domination that I discussed in chapter 1.

Such multiplicities of oppression and domination, though, are also suggestive of a multiplicity of approaches to access issues, and a significant lacuna in our understanding of the socio-political processes of access for people with disabilities relates to the possibilities, indeed probabilities, of significant variations in policy practices. It would also be problematical to reduce access practices to 'all of a type' and, as a range of policy research has indicated, an important aspect of the British welfare state relates to localized discretionary powers which, potentially, permit local authorities to pursue strategies sensitized to localized socio-political structures (Adams, 1994; Healey et al, 1988). Thus, as the chapter will discuss, there is significant local variation in access policies, from authorities which are firmly committed to radical agendas, based on a conception that access is an issue relating to civil liberties, to those which do little or nothing to support access for people with disabilities.

ADVOCACY, ACCESS, AND THE (RE) PRODUCTION OF DISABLISM

As Barnes (1991) cogently argues, institutional discrimination against disabled people is never more obvious than in the restrictions placed on mobility and access by a poorly designed built environment. Throughout the 1970s and 1980s a range of government committees and research reports concurred that access difficulties for disabled people were the 'fundamental cause and manifestation of discrimination' (CORAD, 1982). In particular, the Royal Town Planning Institute (1988) has argued for more sensitive building regulations, 'to reduce the division between the level of accessibility for disabled people and others who are not similarly disadvantaged', yet, while such exhortations have risen since the 1980s, the regulatory framework governing building control, land use and planning control has been relaxed by the abolition of broad standards governing building design and density. This concurs with wider political liberalization programmes pursued by Conservative governments since 1979 which have permitted developers to erect buildings with fewer controls than had hitherto existed. For instance, the abolition of broad standards in housing design has particularly affected special needs groups such as the disabled. Indeed, the Office of Population Censuses and Surveys (OPCS, 1989) show that in the eleven years to 1989 starts of purpose built homes for the disabled fell from 1,129 per year to 230 per year, while for every 100,000 disabled people one wheelchair standard home was started every 13 days (Innovations, 1992).

Part of the difficulty relates to the ineffectual and weak statutory controls governing the access needs of disabled people. As Oliver (1990) notes, the relevant statutes emphasize individual needs, rather than social rights, while the Chronically Sick and Disabled Persons Act (CSDP, 1970), still the most significant statute concerning access provisions for disabled people, reinforces the notion that people who happen to have disabilities are 'people who are helpless, unable to choose for themselves the aids to opportunity they need' (Shearer, 1981, p.82). As chapter 3 intimated, the statutory basis concerning access issues is constructed around a voluntaristic and co-operative, rather than coercive, code, in that the implementation of access provisions is usually dependent on the goodwill of developers. For instance, section 4 of the CSDP Act only requires developers to provide access to buildings where 'practicable and reasonable', while referring only to the provision of access in new buildings, and where 'substantial improvements are made to existing ones'. But, as Barnes notes (1991), the definition of 'substantial' tends to exclude everything which is not a new development, rendering this statutory provision more or less redundant.

The voluntarism inherent in the CSDP Act has also been reinforced by the reluctance of government to legislate on the ground of discrimination and equal rights, exemplified by the Snowdon Committee's (or Silver Jubilee, 1979) recommendation that the provision of adequate facilities for disabled people in public buildings be made a condition of planning consent, an idea rejected by the Labour government of the day. Indeed, CORAD (1982) concluded that statutory frameworks were doing little to 'address the issues', and, as the Silver Jubilee Committee (1979) commented, 'there is no way of compelling a developer to make reasonable provisions if they choose not to do so'. In particular, planning legislation relating to access echoes the CSDP Act in that there has to be a willingness and ability on the part of the local planning authority to inform developers of access standards, as distinct from developers ensuring themselves that they are informed. For instance, the 1971 Town and Country Planning Act, as amended by the 1981 Disabled Persons Act, states that 'planning authorities should draw the attention of developers to the provisions of the 1970 Act and to the BS5810 Code of Practice for access of the disabled to buildings'. This perceived lack of compulsion is also evident in government circulars with, for instance, Welsh Office Circular 21/82, relating to the Disabled Persons Act of 1981, stating that:

> the arrangements for access to buildings can be a planning matter and the arrangements for use by the public, which includes disabled people, raises issues of public amenity which . . . can be material to a planning application . . . conditions may be attached to a grant of planning permission to deal with the matter.

This certainly allows the possibility that access can be a planning issue. Further, Development Control Policy Note 16 (DoE, 1985a) argues for a presumption in favour of considering access:

> when a new building is proposed, or when planning permission is required for the alteration or change of use of an existing building, it will be desirable for the developer to consider the needs of disabled people who might use the building.

None of the above makes disabled access compulsory in the way, say, fire regulations are. While there is clearly an opportunity for interested local authorities to develop access policies on the basis of the above legislation, the system appears to rely entirely on the consent and co-operation of developers. Quite simply, there are no enforcement measures and no incentives which could provide local authorities with something to bargain with.

Moreover, recent planning guidance reinforces the weak status of planning in access matters (a theme which I explore in greater detail in chapter 6). Thus, Planning Policy Guidance 1 (entitled General Principles and Policies, DoE, 1992a) sets out general ground rules on procedures in planning. Following on from Development Control Policy Note 16, it reaffirms the marginality of access, and suggests that local planning authorities should proceed with caution in seeking to enforce access provisions. Thus, as the document states:

Developers and local authorities are encouraged to consider the issue of access at an early stage in the design process . . . Where the public are to have access to the building, the local planning authority should consider the extent to which the securing of provision for disabled people can be justified on planning grounds. The sphere of planning control is limited in this context, conditions attached to planning permissions which have no relevance to planning matters, will be *ultra vires*.

Indeed, as MacDonald (1995), notes, the wording in the document is vague and lacks any sense of how access provisions might be enforced. Moreover, the Access Committee for England (1994a) has argued that it is 'confusing and misleading' while noting that phrases like 'might use', 'are encouraged to', and 'should consider' do little to distinguish what is permissible from what is not. As MacDonald (1995) concludes, 'the negative and vague tone of Planning Policy Guidance 1 is a disappointing start to an examination of disability and planning' (p. 5).

Also, in the context of housing provision, for instance, while local planning authorities were once able to set conditions on the approval of a planning application, often to ensure a mix of housing, including accessible dwellings, such powers have now been superseded by Planning Policy Guidance 3 (PPG 3, DoE, 1992b). As Reeves (1995) has argued, this states that the planning authority can only 'seek to negotiate' accessible dwellings, that access cannot be placed as a mandatory condition of approval of planning applications. This, then, places a high degree of power in the hands of the development industry in determining the levels and types of supply of accessible dwellings. In addition, the DoE (1992b), by continuing to regard accessible housing as somehow something only for those with 'special needs', is sending out signals to developers that only in 'special circumstances' (whatever they are) should such dwellings be provided.[2] The difficulty, of course, is in legally specifying when and where, and in what form(s), 'special circumstances' exist and, as Imrie (1996b) comments, such regulatory provisions reflect government policy which is (at the time of writing) obsessed with the pursuit of utilitarian ideals.[3]

Such ideals underpin the main regulatory provision with regard to access for people with disabilities, that is, Part M of Schedule 1 (1987, updated 1992) of the Building Regulations, a piece of legislation which requires access and facilities for disabled people to be provided in certain classes of new buildings.[4] Possibly because of the technical authority of Building Regulations officers, and the assumption that access is a matter of technical design, of building configurations, most local authorities have relied almost wholly on this part of the planning system to secure their access policy (see Imrie and Wells, 1993). However, critics of Part M point to its proviso that 'reasonable provision be made' in indicating why the regulations are often overlooked and disregarded. That is, provision is not presumed to be mandatory, effective enforcement is difficult, and there are no inducements to encourage better access provision. Moreover, buildings built before 1992 are not covered by Part M unless they have been extended, or converted for a different use and, as Barnes (1991) notes, where buildings are extended, or converted for a different use, the legislation only demands that access should not be made worse.

The Building Regulations also represent the end part of a process which begins with the submission of the planning application. As the next section will illustrate, many access officers feel that it is often much too late to make substantial revisions to buildings by the time planning applications reach the Building Regulations stage. Access provision in the Building Regulations is also limited because they only apply to non-domestic structures, while, even with the 1992 up-dating of Part M, there are few powers requiring owners of established, older, buildings to make them accessible to the range of people with disabilities. Also, as Barnes (1991) notes, there is nothing about listed buildings, while structures are still being erected which are not accessible to people with impairments in their upper limbs or with non-visible impairments (neither are mentioned in the recent revisions to Part M). Moreover, a whole host of institutions are exempt from Part M, like British Rail properties, some educational establishments, crown properties and, most significantly, residential properties, so institutionalizing a partial and inadequate system of access control (seemingly excluding most of the built environment)!

Access policies, particularly in the way that they are defined in and through the building regulations, also reflect societal stereotypes of people with disabilities as being somehow a population who are defined as having mobility impairments. This, as the next section will illustrate, is often translated into practices which largely revolve around creating accessible places primarily for people in wheelchairs. Indeed, time and again planners in practice tend to define 'disability' as analogous to 'wheelchair use'; until 1992, Part M of the building regulation restricted its definition of disability to those with impairments restricting their physical mobility or, by implication, those dependent on a wheelchair (Imrie and Wells, 1993). Yet, as chapter 1 has intimated, wheelchair users are a minority of those with a physical and/or mental impairment. While recent additions to Part M of the Building Regulations (1992) recognize individuals who are visually impaired and/or blind and people with hearing impairments as constituting those with a legitimate physical impairment which might affect their mobility and access requirements, such definitions still fall far short of the much more comprehensive series of inclusions that can be found elsewhere.

Moreover, government advice and guidance on access for the disabled has increasingly emphasized what Oliver (1990) terms a professional and administrative approach to service provision. This is reflected in various pieces of policy advice, including that of the Silver Jubilee Committee which, in 1979, noted that every local authority at district level 'should designate one of their officers, preferably based in the planning department, as an Access Officer who would act as a liaison officer and co-ordinator on questions involving access for disabled people'. Similarly, advice in Development Control Policy Note 16 (DoE, 1985a) and the RTPI (1988) Practice Advice Note 9 indicates that 'local authorities might find it desirable that they should designate one of their staff as an access officer', while government circulars have advised local authorities to facilitate the representation of the interests of disabled people by encouraging the establishment of access groups (C11/85, DoE, 1985b). Such exhortations have been translated into practice and of the 365 local authorities in England in 1994, for example, 259 had designated an individual, either on a full-time or part-time basis, as an access officer (ACE, 1994b).

Yet, while such developments are to be welcomed, for Gooding (1994) the whole voluntaristic system effectively sidelines any meaningful form of participatory involvement by people with disabilities in influencing some of the key discussions on access. Thus, for instance, the enforcement of the CSDP Act rests wholly with the local authorities and it is they, not disabled people, who have the sole powers to bring any actions against owners of buildings who fail to comply with the access regulations. As Gooding (1994) indicates, there is a power imbalance because while, on the one hand, individual citizens with a disability are not legally entitled to appeal against a breach of law by a developer, developers are permitted to appeal against, and contest, decisions on access made by a local authority which they may not agree with. Of course, this is one of the (recognized) weaknesses of the British planning system whereby so-called 'third party' interests are given not much more than marginal rights to contest land use decisions which may affect them. Moreover, it is not at all clear either what access officers, for instance, can achieve or whether they represent not much more than an additive to the procedural and/or bureaucratic structures of state welfarism, a theme I now turn to.

TOWARDS AN APPRAISAL OF ACCESS POLICIES AND PRACTICES

Scotch (1988) characterizes the British system on (in) accessible environments as one of statutes and legal entitlements which provide 'the establishment of a symbolic right to access without substantial guarantees' (p. 164). This observation is reflected in the underlying voluntarism of statutes relating to access and, as chapter 3 intimated, the British approach to disability policy is underpinned by frameworks which continue to perpetuate a professional-client divide which reinforces the idea that people with disabilities are dependent, unable and, as such, require the (extended) services of the welfare state to enable them to gain a measure of social equity. For Young (1990) the institutionalization of state practices, like access, in this way propagates such ideological propositions as a self-serving mechanism to maintain the divisions between the enablers and the dependent and, as Young rightly notes, such systems are bereft of participatory democracy (a theme which I address directly in chapter 7).

Others, like Oliver (1990), situate such developments in an advocacy model, which, while espousing the integration of the disabled into society, reinforces the importance of professionals and the existing socio-political and legal frameworks in providing solutions for disabled people. In particular, the institutionalization of access really reflects the wider broadcloth whereby public policy and welfare services have become increasingly de-politicized. As chapter 3 has intimated, an important part of the welfare approach to disability has been the constant striving to present it as a 'distributive' issue where, as Young (1990) notes, 'background issues of the organization of goals of production, the positions and procedures of decision making, and other such institutional issues do not come into question' (p. 70). Indeed, while many planners and/or access officers do not like the strictures of the legislation on access, most, if not all, choose to use it to organize their policies for accessible environments (Reeves, 1995). As the evidence presented in the next section will indicate, most access officers also tend to support the idea of 'design' solutions to inaccessible places, while medical definitions of disability are much more prevalent than social constructivist ones. For many, their position, as 'an expert', is also seen as inviolate and a condition of any measurable successes in achieving accessibility for people with disabilities.

Yet access officers are clearly constrained by wider structures and it is not surprising that their practices are highly proscribed by statute and they avoid risk by broadly following the voluntaristic tenets of Part M of the Building Regulations (Barnes, 1991; Imrie and Wells, 1993). It is also clear that practices vary considerably between local authorities, yet there is little available detail about many of the crucial roles, responsibilities and problems of access officers in seeking to secure accessible environments. This, then, is the empirical focus for the rest of the chapter in which I explore critical elements of the access officer's role in seeking to create accessible environments for people with disabilities. In doing so, I draw on a range of empirical data sources including semi-structured interviews with access officers and/or access officers designate in twenty-two local authorities in England and Wales (referred to as the interview survey). The authorities ranged from rural to urban, Conservative controlled to Labour, and from those with highly politicized disability pressure groups to those with weakly developed ones. Interviews took place over the period from January 1992 to March 1995. Additional materials were also gathered, including local plans, supplementary planning guidance on access, design guides, minutes of meetings between access officers and local access groups.

In addition, a postal questionnaire (hereafter referred to as the postal survey) was sent to all the English local authorities who, in March 1994, had an access officer (see appendix 1). The list of officers was obtained from the Access Committee for England, giving a total of 259 authorities. One hundred and twenty questionnaires were returned, a response rate of 46 per cent. In combination, the two surveys provide significant information on procedures and policies of access. As the evidence will indicate, there are considerable variations in access practices and in the values, attitudes and forms of support of local authorities. I divide the discussion into two. First, I discuss the wider structural conditions within which access practices are derived, from how local authorities define disability to the levels of fiscal and political support that they give to access initiatives. In a second part, I consider the specific roles of access officers and I provide an assessment, often based on self testimony, of how far, and in what ways, they are able to address the (perceived) needs and concerns of people with disabilities. In order to preserve the anonymity of interviewees, the text does not name individual people or places when discussing their views. Qualitative material in the form of quotations used in the text are derived from both the postal and the interview surveys.

Institutionalizing access

Since the Silver Jubilee Committee (1979) the development of access policies in Britain has gathered some momentum, to the point whereby 259 local authorities in England had a designated access officer in 1994, while most localities have one or more disabled access groups seeking to influence local planning policies and procedures. Indeed, as the Access Committee for England (1995) has indicated, most local planning authorities have some form of access policy whereas, ten years previously, little or nothing existed. For example, 74 (62 per cent) of respondents in the postal survey had no adopted policy on disability and access. Of those which had a policy (46 or 38 per cent) the majority, as Table 5.1 shows, have only recently adopted it. Such policy frameworks, however, are characterized by a vagueness in terms of how disability should be defined and who the target population for the policy is.

Table 5.1 Date of adoption by sample Local Authorities of a formal policy on access

Date of Adoption	Numbers (%) of Authorities Adopting
pre 1990	10 (21.7)
1990	4 (8.6)
1991	4 (8.6)
1992	8 (17.2)
1993	11 (23.9)
1994	8 (17.2)
1995	1 (2.0)

Source: Postal Survey, 1994–95

Thus, a strategic absence in many of the postal and interview sample authorities is an understanding of who it is they are creating accessible environments for. This was, in part, reaffirmed by the postal survey in that 74 (62 per cent) of the respondents had no working definition of disability.[5] Of those that claimed to have some form of definition (42 or 36 per cent), over half referred to Part M of the Building Regulations which, as I discussed in chapter 3, provides a reductionist and partial, thus problematical, definition of disability. For the rest, the definition of disability tends to be a mixture of the technical and the apolitical, with only three authorities providing some evidence of seeking to place disability in a wider, critical, social context. For example, as one authority, from the postal survey, commented, 'disability is oppression of people with impairments. Being a disabled person is defined by each individual so if you say you are disabled then you are. The definition of disability is a social one not a medical model.' Such sentiments were echoed by another postal survey authority in reflecting a social constructivist perspective: 'people have impairments but it is society that disables them. With the right planning in our environment we can enable rather than disable.'

Such definitions were exceptions and much more typical was the deployment of medical analogies and/or metaphors, so reinforcing the notion that somehow disability is a problem of and for the disabled person. One postal survey authority, for instance, was unequivocal in stating that 'the medical model of disability is the one we chose to use . . . it's the most appropriate . . .'. Others reinforced the notion of individual impairment as the source of the problem facing people with disabilities and common definitions of disability were 'those with physical mobility deficiencies', as though the person's deficient mobility is a problem wholly explicable in terms of their impairment. In addition, one authority, a large city council, deployed the most common conception in defining disabled people as those 'with a physical impairment which limits their ability to walk and people who need a wheelchair for mobility'. One of their draft documents on access qualifies this with the contestable statement that, 'this focusing on the problems of wheelchair users can be justified because this group often faces the greatest physical barriers in buildings'. This, then, reinforces the problematical stereotype that disability is reducible to a specific form of impairment which policy, in turn, should plan for.

Such definitional differences are important because they, in part, inform the types of policies which are being pursued by local authorities. Thus, two major metropolitan authorities, in using social constructivist definitions, have interlinked access policies with the pursuit of equal opportunities for people with disabilities. In doing so, they are being proactive in developing wide-ranging initiatives which seek to politicize disabling environments as part of wider structural forms of discrimination (although neither referred to the term oppression as a dimension of the experiences of people with disabilities). As one of the access officers noted, 'I'm influenced by the local radicalism in the city towards disability and it's always been part of an equal opportunities agenda . . . the council want it this way and so do I . . . so, we're involved in supporting campaigning for civil rights and we fund disabled people to go and protest at appropriate times'. Yet, in contrast, others see access as the implementation of the core legislative provisions, and of adhering to the Circular advice of the DoE. As one authority commented, 'our hands are really tied and we can't do much more than is prescribed for us in the statutes . . . that's the way we proceed'.

In part, this seems suggestive of a minimalist approach to access, yet, from the interviews with local authorities, all but three had produced policy statements and/or policy leaflets concerning access for the disabled. While the majority felt that access issues were more significant than five years ago, in that they were much more to the fore of policy agendas, most commented that access is still marginal to the main work of the authority. As one authority argued, 'we neither have the time or the resources to get that involved with access matters'. Moreover, the majority of the sample authorities were in environments that could not be described as 'user-friendly' to disabled people. While many of the places I visited had some recent refurbishments, most were typified by few facilities for the disabled and little fore-thought about enhancing accessibility. Moreover, no single authority had any knowledge of the actual demand for a more accessible environment, and most tended to define 'the disabled' as wheelchair users. As one authority commented, 'we don't know what percentage of people are disabled, we don't know how many blind or deaf there are, we have no information and there's little pressure put on us to provide it'.

In terms of securing access gains for people with disabilities, a whole range of opportunities and problems is evident. As Table 5.2 indicates, the types of restraints on access officers, in seeking to secure access provisions, are wide-ranging. For most, a combination of wider, structural, restraints was identified as of importance, from the lack of finance to pursue access projects to the poorly conceived nature of the statutes which do little to help them enforce access policies. As some of the respondents commented, a significant difficulty relates to the lack of clarity of legislation or, as one respondent noted, 'there is a real lack of government advice and guidance for determining whether or not access is a material consideration when we consider planning applications by developers . . . it's unclear and ambiguous'. For most authorities, the post of access officer brings little or no funds with it and even where full time access officers have been appointed the allocated budgets do not extend much beyond paying for the salary. Most respondents also mentioned 'lack of time to represent access issues' as an issue, yet this is hardly surprising given, as the next section will discuss, the predominantly part-time nature of the role.

In particular, the issue of the resourcing of access was raised time and again and, as one full time access officer, in interview, commented, 'the reason for the shortage of resource in this area is the myth that a generally sharpened awareness of problems of access is itself enough to ensure proper practice'. Yet the funding for the officer, derived both from the Equalities Unit and the Planning Department, was providing a salary and, in 1994, a budget of £20,000 for access initiatives like educational programmes, workshops, promotional activities, and training schemes and was allocated far more than for any of the other sample authorities. Indeed, of the twenty-two authorities interviewed, only three, all with full time access officers, had been allocated a budget over and beyond the salary. In contrast, the observation of one of the part-time officers is apt in that 'I don't have any resource to do anything with . . . so I can't do much . . . it's all low priority stuff really'. For others, resources were being culled inter-departmentally, and, as a full time access officer from a

Table 5.2   The main problems facing access officers in securing access gains for disabled people

Type of problem (Total responses)	A significant problem Numbers (%)	A problem Numbers (%)	Not a significant problem Numbers (%)	No problem Numbers (%)
Lack of time to represent access issues (108)	31 (25.7)	38 (35.1)	27 (25.0)	12 (11.1)
Lack of finance and resources (111)	55 (49.5)	33 (29.7)	18 (16.2)	5 (4.5)
Indifference amongst planning officers (108)	7 (6.4)	26 (17.5)	51 (47.2)	24 (22.2)
Lack of support from councillors (108)	7 (6.4)	19 (17.5)	49 (45.3)	33 (30.5)
Lack of powers to enforce access policies (108)	34 (31.4)	44 (40.7)	26 (24.0)	4 (3.7)
Resistance from private developers (108)	36 (33.3)	47 (43.5)	22 (20.3)	3 (2.7)

Source: Postal Survey, 1994-95

London borough noted, 'I have no budget and I have to go to different departments and try and tap their budgets . . . so, if planning has a budget for pedestrian schemes, or, say, creating a riverside walk, I make sure some of this is allocated to access'.

Such strategies, in part, reflect the more general squeeze on local authority budgets over a period where central government spending targets have sought to reduce expenditures. In addition, as one interviewee noted, 'it's also because access is quite new and we're having to fight to establish its credibility with the more established spending programmes'. Others noted how they are often involved in a scramble for resources, yet never get too far. As one interviewee commented, 'there's a common feeling here that access is a soft subject and really shouldn't be resourced at all . . . there is some officer hostility . . . I've got to be really pushy to get anything'. For this individual, being pushy had extended to him tapping Urban Programme funds while also slotting into a recently awarded City Grant scheme.[6] And, as he concluded, 'I'm now negotiating with development teams about prior allocation of finance for access, although they all want to see it as additive, as an extra, while I'm really battling to persuade them that access is integral to any scheme . . . I'll get there though'.

Moreover, an important part of 'creating accessible environments' is the attitude of private sector developers and their willingness (or not) to comply with access regulations. As one access officer commented, 'they always complain about the costs of putting in the additional access features but most never really know what it will really cost them if they have to do it'. Indeed, of the 120 postal survey returns, 76 authorities (or 65 per cent) concurred that private sector developers need some persuasion but will accede to requests and, as one authority noted, 'the situation is really variable but the general pattern is predictable. If it's statutory, they accede, or if it's council owned land they will agree.' Another stated, 'our attitude to the issue is to tell the developer that an accessible building is one which maximizes their market opportunities.' Indeed, this seemed to be a common, binding, theme with most authorities commenting on the cynicism of the development industry, able and willing to provide appropriate design features for disabled persons if it presents them with a market opportunity. In one example, a developer approached the planning authority to obtain advice on how to convert their Welfare Club to cater for disabled groups who were seeking to rent out a place for their functions. As the respondent rationalized it, the Welfare Club were only willing to make the required changes to their premises because 'money was there to be made out of it'.

While the majority of new public buildings in the sample authorities have wheelchair access, one authority provided an illuminating rationale underpinning access to their building. As the interviewee commented, the construction of ramps into the building was justified on the grounds of letting 'women with pushchairs into the building to pay their poll taxes', not with any disability or access concerns as the prime motivator. Moreover, the local authority offices in this authority are a ten minute walk from the nearest carpark, hardly conducive to giving disabled people access to the 'centres of power'. The authority, like most others, also noted the absence of tactile walkways to enable the visually impaired to enter the building, while commenting that none of their staff knew any sign language to communicate with hard-of-hearing people. Likewise, another authority commented on their credibility in requesting developers to adhere to access regulations. According to an officer, their public buildings are disgraceful, in that not a single one has a disabled toilet nor ramp access. Indeed, the disabled are wholly prevented from gaining access to the actual council chambers in the borough and, as the officer commented, 'it is difficult to preach to developers about accessibility when we're wholly inaccessible ourselves'.

Most authorities also agreed that the level of councillor awareness and support of access issues was a crucial feature of the local policy environment. As one authority suggested, 'disability issues are not a vote catcher down here' while another noted that there is a 'reluctance to spending here and there haven't been any debates about access issues in the council', concluding that this was not really suprising given that the disabled 'can't even get access to the council chambers here'. In contrast, one of the city authorities noted the commitment of one councillor in the setting-up of the local access group, a process enabled by the general commitment of the local council towards equal opportunity policies. Indeed, the majority of interviewees were able to identify key local politicians who were, as one expressed it, 'indispensable to raising local awareness of disability issues', even if, single-handedly, they were unable to effect significant change. In concurring with this, as Table 5.2 indicates, most of the postal survey respondents felt that councillors pose little or no problem in the pursuit of access and, as one authority noted, 'we've got some very supportive local politicians . . . they make all the difference' .

However, for most of the sample authorities, both postal and interview, access issues were a low priority, attracting few funds or resources, with few examples of separate budgets with which to manage or develop access policies. As one authority concluded, 'we're too busy getting on with the normal workload to be bothered with additional tasks'. Moreover, the reliance on Part M of the Building Regulations, as the definitional basis of 'disability', while convenient for local authority officers, is problematical because it reinforces the notion that somehow accessibility is reducible, and soluble, through forms of technical (design) control. In turn, this provides a weak basis for the politicization of access issues within local authorities, a situation compounded by the marginal roles and weak status of access groups (which I discuss more fully in chapter 7). In addition, most authorities were resigned to letting the market 'facilitate access', while developing and supporting schemes which, while alleviating some of the symptoms of disabling environments (e.g. the removal of obstructive street furniture, placing tactile walkways in pedestrian schemes, etc.), were falling far short of a significant transformation of the built environment.

Access officers: an enabling role?

Creating and sustaining a momentum behind the implementation of access legislation at a local level increasingly rests with a designated individual in the local authority, the access officer. As already indicated, various government circulars and advice notes highlight the potential and possibilities of access officers in representing the disabled. Indeed, as surveys on access, conducted by the Welsh Council for the Disabled (WCD, 1987, 1990) have noted, despite the growing body of official and semi-official advice on the desirability of local authorities appointing or designating an access officer, few authorities in Wales have seriously considered the matter (1987, p. 18). Of the authorities I interviewed all but six had part-time officers (five had a full-time officer, the other no officer at all), of which the majority were employed full-time either as building control or local planning officers and few had a physical and/or mental impairment (see Table 5.3). Moreover, as Table 5.3 shows, the majority of posts have been set up since the WCD survey, indicating that access, as an issue, has proliferated in recent years. This data concurs with the postal survey and, as Table 5.4 shows, of the 120 sample authorities only 18 were employing access officers on a full time basis.

In line with the WCD (1990) research, both of the surveys indicate that designated, part-time, officers tend to operate either in Building Control or Planning departments. It is still assumed that the proper place for the access officer is in Building Control, although one authority argued against this as 'locking the stable after the horse has bolted', referring to how the 'form of most developments are already agreed on by this stage'. As Table 5.3 shows, a

Table 5.3 Characteristics of designated access officers

Local Authority	Designated officers' normal duties	Date of appointment of first access officer	Does the access officer have a physical and/or mental impairment
A	Planner, LP	1989	No
B	Full-time AO	1990	No
C	Planner, DC	1990	No
D	Officer, DC	1990	No
E	No designated person	–	–
F	Deputy Chief Planner	1985	No
G	Planner, LP	1991	No
H	No designated person	–	–
I	Planner, LP	1991	No
J	Full Time AO	1987	Yes
K	Full Time AO	1989	No
L	Full Time AO	1990	Yes
M	Full Time AO	1991	Yes
N	Planner, DC	1990	No
O	Officer, DC	1991	No
P	Officer, DC	1993	No
Q	No designated person	–	–
R	Officer, DC	1994	No
S	Officer, DC	1989	No
T	Deputy Chief Planner	1986	Yes
U	Officer, DC	1992	No
V	No designated person	–	–

AO = Access Officer   DC = Development Control   LP = Local Plans

Source: Interview Surveys, 1992-95

Table 5.4   Status of access officers in local authorities

Status of Access Officer	Numbers (%) Appointed
Full Time Appointment	18 (15.5)
Part-Time Appointment	84 (69.0)
No Appointment	18 (15.5)
Total	120 (100.0)

Source: Postal Survey, 1994-95.

high proportion of the authorities interviewed had placed access duties in Building Control, a situation confirmed by the postal survey, with over half of the access officers located in Building Control departments. Moreover, because the majority of access officers have responsibilities elsewhere, the time that they devote to access is limited. Thus, as the postal survey indicates, the proportion of time that part-time access officers devote to access duties is usually considerably less than a quarter of their allotted work hours, while their access duties tend to be limited to two or three key tasks (Table 5.5). This concurs with the WCD (1990) survey in that access officers tend to devote most of their allotted time to community consultation, meeting access groups and going through planning applications with them. Indeed, there was little evidence from my surveys of positive, proactive, work being undertaken; positive initiatives, programmes and policies are beyond their remit and time budgets.

Table 5.5   Time expended on access duties by designated access officers

Time spent on access duties (by %)	Designated access officers Numbers (%)
Less than 5	34 (33.3)
6 to 20	34 (33.3)
21 to 50	16 (15.6)
100	18 (17.8)

Source: Postal Survey, 1994-95.

In addition, what tends to rule out a full-time post is the cost and most authorities admitted that a full-time officer might have been appointed in less austere times. Yet, as interview material suggests, the role of part-time officers tends to be ill defined, ambiguous, with duties confined to reactive measures often with little time or resources to develop active access policies. In these organizational senses, a part-time access officer is no panacea for the deficiency of active access measures for the disabled. For instance, in one case, access was dealt with by somebody who happened to have a personal interest in access issues although, when she left the authority in 1987, these functions were not passed on to another officer. In another case, the 'designated' access officer felt that her role was ill defined and that, theoretically, her access duties should be more fully linked into the policy units of the authority. The lack of any clear guidance made her job difficult to perform, while the pressures to prioritize her 'normal' duties in development control made it impossible to give up any more than two hours a week on access duties. This, then, highlights a range of issues concerning access and the planning process. Foremost, the ill-defined nature of the access role tends to lead to the designation of an officer by default, rather than by active policy considerations. This tends to reinforce a perception that 'access is not important' and residual to the main work of the department.

Moreover, because the designated officers tend to work in (and from) specific departments, the job often becomes more closely associated with a series of single functions of the authority, rather than be seen as integral to the multiple functions of the local authority. Typically, this may mean, for example, that the building control officer is also the access officer, and that access issues are simply, as one respondent put it, 'framed within the day-to-day functions of the department you work for'. In this context, access might be something reducible to building control matters. In addition, the diversity of local response to the question of who fulfils the access officer's role reflects the lack of national legislation and standards to support a more systematic and comprehensive provision. A typical pattern in local authorities is that access issues tend to be highlighted through the endeavours of particular individuals, councillors, officers or members of the public, people who have a specific interest or commitment to disability rights issues. Indeed, without this clear framework, and without the resources to back-up the appointment of full-time officers, one authority commented that full-time officers would 'provide a point of contact for pressure groups, although it'll give them an expectation of actions which can't be delivered due to political apathy and a lack of money'.

At the departmental level, interviewees generally concurred that there is little awareness of disability issues amongst officers, and even less time to think about them. As one authority commented, 'developing any awareness depends on having the time to do so . . . our responses to access issues are of the fire-brigade type'. In particular, it seemed that a significant minority of authorities had allocated access duties to officers almost by default, with an attitude that the access job really gets in the way of the 'real' work of the department. For instance, in one authority, no one seemed to have recognizable access officer responsibilities, although an individual officer, only recently appointed, had by default (i.e. no one else wanting to do the job) acquired duties to liaise with the local access group. As she indicated, 'I've only been here two months and it seemed I was the convenient person to give it to . . . it's as though the job was passed to me as I was new and they knew I wouldn't (couldn't) create any fuss',

Such examples contrast with authorities where the role of a designated access person is deemed to be wider than a single person. In one of the authorities, for example, a collective approach is taken in dealing with access issues, and a more proactive approach by officers is evident. The approach of this authority is illuminating in that all planning officers are expected to take responsibility for highlighting, discussing and incorporating access issues into various parts of the planning process. In particular, officers with responsibilities for development control in different segments of the locality regularly meet with the access group, thus providing a high level of officer input into access issues. As the authority noted, 'we prefer maximum officer involvement . . . this is one of the best ways to educate our officers in access issues, to put them all into the firing line'. Other authorities were also critical of creating a separate access officer role and, as a respondent from a large city authority commented, 'my access duties are listed twenty fifth in my job description because the council feels we should all be aware of the issues . . . they're not too happy with this figure head role' .

However, such ideals demand a responsive authority able and willing to inculcate values in local authority officers sensitized to access. In practice, I found wide variation in the amount of training given to support access with designated access officers usually required to learn on-the-job. In one case it was admitted that a technically proficient access officer would be an improvement, given that the present incumbent felt that he did not really understand Part M of the Building Regulations nor the technical issues associated with creating more 'accessible environments'. This view was reinforced by another authority who concurred that no training had been given to the designated officer, while the only way to learn about the issue was to pick up any available literature on planning and access. There is also evidence of initial officer resistance to training in access and, as one full-time access officer noted, 'I encountered resistance, and even hostility, from departments on taking up the appointment'. Part of this was connected to the officer's lack of identity within any one department, and also to the attitude of Building Control that 'I was doing their job'. As the access officer concluded, 'I had to put myself about to begin with, but now everything is OK . . . I spend a lot of time trying to educate the officers about the issues and I liaise with lots of local groups'. However, the officer did conclude that there was a lack of resources and that 'I still have a long way to go to convince people that the position isn't a charitable event'.

In contrast, the authorities with full-time officers were generally active in developing and initiating new policies for the disabled – for example, one city authority produced design guidance notes for developers and organized seminars for architects and developers. The officer has also produced an umbrella policy for the whole district, a policy which was accepted by the council in November 1990. This commits the authority to 'promote equality of opportunity for disabled people' while encouraging 'people and agencies to provide improved access to buildings and information for both disabled and able-bodied persons'. Moreover, the access officer has succeeded in persuading the Estates Committee to allocate five per cent of city centre car parking spaces for disabled users. The local authority has committed a substantial portion of funds to access policies, including a new Pedestrian Priority Area which has been designed to incorporate state-of-the-art features for people with a variety of disabilities (including tactile walkways for the blind and partially sighted, and 'user-friendly' street furniture). In this sense, the possibilities for access are greater than many planning authorities acknowledge and/or are aware of and a range of initiatives suggests that there are few environments which cannot be made accessible to people with disabilities.[7]

One of the crucial elements, however, in seeking to generate accessible environments relates to the interface between the policy institutions and people with disabilities. While chapter 7 explores this in some detail, it is clear that the institutionalization of access is operating through clearly identifiable groups, particularly access groups which are usually, in principle, operated for and by disabled people. Yet access groups, while variable in form, resources and policy objectives are characterized by bounded powers within local authority structures and very few have access to the more powerful policy and resources committees (ACE, 1994c). As one access officer commented, 'I'm here to convey my expertise to them and to help them through the minefield of legislation and statutes'. Such paternalism was evident amongst a range of interviewees and, in those authorities where there was an access group (in fifteen of the twenty-two authorities) the overwhelming forms of interaction between them and their local authority were of a consultative and/or advisory nature. As an access officer noted, 'they inspect the planning application forms and get to comment on any major schemes . . . we listen to them and take their views into account'. Such voluntarism, in turn, is overlaid with what local authorities seem to regard as the legitimate representatives of disabled people. As chapter 7 will discuss, most respondents were unequivocal in stating that they would never deal with anyone outside of the formally constituted (access) group.

CONCLUSIONS

Systems of access representation, while going some way towards acknowledging the inequitable and oppressed situation of people with disabilities, are characterized by structures which do little to redress the problems of access in the built environment facing people with disabilities (Imrie, 1996b). Research indicates that access issues are still marginal to the main work of local authorities, a situation reflected in the low levels of resources devoted to them. In particular, most local authorities tend to develop access duties as an 'afterthought', allocating the duties to an officer who receives them almost by default. This is unsatisfactory because not enough time or training allowance is made to permit the officers to discharge their access duties. Inevitably, under pressure, one of the first jobs to be put to one side is the access responsibilities. However, the idea of a figure-head, with sole responsibility for access issues, is also problematical given the high level of ignorance amongst most planning officers concerning access issues. As one authority commented, 'the idea of a specific post to oversee access issues seems ludicrous when every planning officer should be versed in the relevant legislation, issues and debates to permit them to discharge access duties as and when appropriate'.

Moreover, creating accessible environments is more or less impossible in a context where the framing legislation is weak, reactive and capable of doing little to challenge the practices of private developers who maintain, time and again, that the costs of an accessible environment are prohibitive. This is a message which successive governments have upheld and limits the role and effectiveness of access officers in a number of ways. In particular, (legitimate) discussion and decision making takes place within a narrow range of distributive issues such as, for example, how much funds are required to create accessible shops, should funds be allocated for a housing project or a pedestrianization scheme, and/or will access schemes create jobs. As Young (1990) argues, such discourses suggest that the ends of governance are already given in that 'policy has always been oriented to the best way to allocate the surplus for individual and collective consumption rather than the more central question of the best way to control the power to realize social needs and the full potentialities of human beings' (quoted in Young, 1990, p. 71, from Smith and Judd, 1984, p. 184). In the context of access, then, the general approach seems to be about shuffling limited resources around with little by way of a wider politicization of the process. This discounts awkward questions being asked, like how and why people with disabilities are disadvantaged, both by policy practices and wider socio-economic processes.

The prescribed role of access groups is also of significance, and research seems to suggest that they are 'managed' by the local authorities that they interface with. While not denying particular layers of autonomy that such groups possess, the interest group politics, of the type that the systems of access seek to propagate, has the effect, as Young (1990) argues, of locking 'individual citizens out of direct participation in decision making and often keeps them ignorant of the proposals deliberated and the decisions made' (p. 73) (also see chapter 7). For the most part, pluralist politics fails to recognize the 'individual' person and people are only permitted to voice their concerns through the context of specific government programs and/or formally constituted institutions. In this sense, there is an effective decoupling process or one in which, as Young describes it, a particular individual's self interest becomes 'incoherent' (p. 73). In the context of the access officers' multiple roles, it is clear that their dealings are wholly defined in terms of 'interest group pluralism' and, as one access officer described in interview, 'we'll only deal with the recognized groups, we haven't got time to do any more than this and anyway this is the only way to get anything done'.

NOTES

1. The 'access industry' includes national co-ordinating committees like the Welsh Council for the Disabled and the Access Committee for England which was set up in

1983. The Centre for Accessible Environments, based in London, is a registered charity and is one of the main organizations for bringing together planners, architects and people with disabilities to discuss issues relating to access.

2. Reeves (1995) comments that there are challenges to this official view, propagating the idea that all housing should be inclusive of the highest accessibility standards thus allowing access for all. One of the more powerful statements is contained in the European Manual for the Accessible Environment (Langton-Lockton, 1994). As Reeves notes, 'the philosophy . . . is that it is normal to have a disability and that those involved in the built environment should design and plan inclusively rather than tacking on supplementary provision for people with disabilities and other special needs groups' (p. 5). However, as Reeves reports, few local planning authorities challenge the official view.

3. The reality for people with disabilities has been worsened by government housing policy. This favours home ownership and public housing building schemes have diminished in recent times, while the collapse in house prices after 1989 has led to house repossessions as people default on their mortgages. Also, organisations such as housing associations, which have traditionally tried to provide housing for people in need, are increasingly being forced to charge 'market' rents in order to off-set the diminishing levels of financial support they receive from government. Such situations are compounding the general lack of affordable housing in the UK and the poor are those who are affected most. Morris (1992) estimates, for instance, that there was a 92 per cent increase in homelessness amongst people with disabilities from 1980–88 compared with 57 per cent for all households.

Such situations are compounded by a range of organizations. For instance, the attitude of the House Builders Federation (HBF, 1995), a national campaigning organization for British housebuilders, towards access is revealing about the difficulties facing access officers and/or planners in trying to gain accessible housing for people with disabilities. The HBF's response to the government's recent proposals to extend the Building Regulations to new domestic dwellings has been hostile. The HBF say, for instance, that 'we do not consider that the problems of access to new homes for the extremely small numbers of wheelchair users likely to visit one are particularly severe'. Yet, one must ask what their attitude would be if it were they who were in a wheelchair. Their dismissive, crude, even ignorant views are revealed throughout a document that is wholly concerned about costs to housebuilders, yet nowhere do they support their claims that the effect of the new, proposed, regulations 'will be to add to the cost of dwellings in key sectors of the market'. The HBF are also disablist by associating the proposed regulations with taking 'numbers of attractive and popular housetypes out of the market' as though designing for people with disabilities is necessarily producing 'unattractive and unpopular' houses. Their implied association is offensive.

4. Part M of the Building Regulations (England and Wales) relates to access into and around buildings. It was established in 1987, extended in 1991, and is set to be extended again. It only requires access when new developments occur and/or where substantial redevelopments are taking place. The extended provisions (1991) require access to all floors, yet its major weakness, as I have highlighted elsewhere, is that housing is excluded from the regulations. Contrasts with other countries are quite illuminating. Thus, while people with disabilities are still treated like second class citizens in places like Sweden, Swedish building codes require accessibility in all new multi family housing. Thus, apartments built after 1977 will have no steps leading into them and apartment blocks of three storeys or more must provide lift facilities. Yet for

older housing the access codes are less strong and local authorities have a high level of discretion as to what extent an existing structure has to be made accessible and cost is the determining factor.

5. There are exceptions to this. For instance, research commissioned by York City Council (1993) shows how prevalent disability (following the World Health Organisation definition) is. The research indicates that, in 1993, there were '14,000 people with one or more disabilities, which is just under 14 per cent of the population. 70 per cent of these are aged 60 or more, emphasizing the age related nature of disability. Overall, the most common area of disability was locomotion (70 per cent) followed by hearing (42 per cent). Approximately 1 in 60 are blind or partially sighted' (p. iii).

6. Since 1981 British urban policy has been underpinned by a range of initiatives. Its mainstay was the recently abolished Urban Programme (UP), set up in 1968 to target resources at any place that could demonstrate 'special social need'. The Programme involved positive discrimination and was characterized by small scale projects, many carried out by voluntary groups. It was focused on 57 different places in the United Kingdom. The social-welfare approach of the UP was gradually eroded during the 1980s by the introduction of a range of schemes which were increasingly concerned with revitalizing the inner cities by encouraging property development. One such scheme was City Challenge (CC) established in 1991 to counter some of the criticisms that communities were being excluded from participation in urban policy programmes. To qualify for funding under CC, local authorities enter rounds of competitive bidding, presenting project proposals to the Department of the Environment that have to demonstrate a commitment to local partnerships. To date, there have been two rounds of bidding and 31 authorities have received funding.

7. One recent scheme, for instance, involves Leicestershire County Council in a partnership with the Countryside Commission and Scope in the design of Britain's first woodland for people with disabilities, incorporating features like tarmacked circular paths with gradients of less than 1:20, picnic and play areas with specially designed play equipment, and parking for disabled vehicles.

FURTHER READING

Scotch (1988) provides some useful insights into access issues, yet little has been written about this topic. The offices of the Access Committee for England is worth visiting to gain copies of current research reports while the only research published on access practices in the UK is Imrie and Wells (1993). Barnes (1991) provides some useful general overviews on access and the built environment.

6

Institutional Mediation and Planning for Access

INTRODUCTION

One of the most important institutional contexts for dealing with disability and access in the built environment is the environmental planning system. As the previous chapter has intimated, land use planning is at the core of the decisions concerning access into and around buildings yet our understanding of how such systems intervene in, and influence, the qualitative dimensions of access is limited (although see Imrie, 1996c; Imrie and Wells, 1992, 1993; Reeves, 1995). Indeed, for many planners in practice access, until recently, was always seen to be more of a concern for social welfare or other policy departments, and even where it was considered to be a planning matter the particular discourses which were utilized tended to reduce questions of access to ones of the 'appropriate technical standards'. In this sense the technocratic ideology, which has long been a pervasive influence in underpinning the 'identities' of planning practitioners, was, and in some senses continues to be, an important way of denying, even mystifying, the socio-political content of access issues.[1] This, as Giddens (1991) has documented, was often a self-serving way of maintaining the guise of a political neutrality, of, essentially, depoliticizing the practices or procedures of planning from its wider substantive contexts.

Such manoeuvres by the planning profession were, of course, linked to wider concerns about legitimacy and status, yet the inability for such conceptions to be maintained is illustrated by the range of evidence which clearly indicates that planning theories and practices are not neutral either in terms of their socio-political origins or in their broader distributive effects on communities (see Healey et al, 1988). Indeed, spatial development policies, post 1945, have been implicated in wider processes of uneven development in the cities (Adams, 1994; Ambrose, 1994; Healey et al, 1988; Thornley, 1991). A range of evidence suggests that planning policies are imbued with dominant socio-culturalist conceptions, comprising a gender bias and colour blindness while supporting land use developments which demonstrably reinforce a class differentiation in terms of spatial distributions (Gleeson, 1995; Healey, et al, 1988; Imrie, 1996b; Thomas, 1995). Thus, as Thomas

(1995) notes, for example, in relation to race, planning ideas have long sustained and reproduced ideologies which have contributed to the poorer material conditions of Britain's ethnic minorities; and, as Laws (1994a, 1994b) has documented, in relation to planning for the elderly, it is hardly surprising that built environments are characterized by their ghettoization given the dominance of ageism as one of the core oppressive values of western societies.

Likewise, it seems clear that some of the dominant ideas which have underpinned planning practice have been far from sensitized to the access requirements of disabled people. For instance, many planning ideas, theories and practices emphasize mobility over accessibility and, in doing so, inadvertently discriminate against mobility impaired people. Thus, for example, the emergence of the post war planning ideal of the UK New Towns, combined with policies to minimize development in areas of protected green belt, was implicated in the spatial extension of urban structures while placing a premium on individual mobility to gain access to facilities which had previously been spatially constrained. Similarly, zoning ordinances, in both the Canadian and the USA planning systems, have been used as self conscious tools with which to construct 'special' environments for groups like the elderly and the disabled, effectively taking them out from the mainstream (see Laws, 1994b, for an instructive account of this). Such illustrations, in part, indicate the possibilities that environmental planning systems, while attentive to certain aspects of disabled access may, more generally, be implicated in reinforcing and (re) producing disablist built environments.

However, few studies have addressed the socio-distributional impacts of contemporary environmental planning policies on the lives of people with disabilities while there is little or no research to show how planning systems might be implicated in the (re) production of disablist socio-spatial structures (although possible exceptions are Imrie and Wells, 1992, 1993). Given this proposition, the chapter seeks to explore the role and nature of planning policies and practices in shaping, reinforcing and challenging the tenets of disablism in the city. Based on both primary and secondary research data of practices in the British planning system, the chapter develops the contention that planners do have the capacities and powers to influence the physical nature of the built environment although they have limited influence over the derivation and perpetuation of such (disablist) spaces and places. In charting the practices of planners, the chapter documents the critical utilization of the key tools of the planning system, that is, of the role of the land use plan in addressing access and disability, while discussing how far, and in what contexts, planning control systems are utilized (or not) to modify developments in line with access directives.[2]

NEO-LIBERALISM AND THE EMERGENT PLANNING SYSTEM: THE IMPLICATIONS FOR DISABILITY AND ACCESS

One of the more protracted silences in the planning literature relates to the relationship between planning practices and policies and the multiple forms of disadvantage and oppression which people with disabilities have to confront (although see Imrie, 1996b; Thomas, 1992). The orthodoxy seems to suggest that there is no significant relationship between the two, that somehow the institutions and practices of planning do little to reinforce the weak, marginal and dependent status of disabled people. Such conceptions are supported by the complicity of the planning schools which, until recently, taught little or nothing about equal opportunities and certainly ignored disability and access issues. Indeed, most planning schools still ignore disability as an issue and what is taught tends to be treated as a 'special subject' seemingly peripheral to the main task of inculcating trainee planners with the spirit and ethics of a technocratic ideology. Likewise, there are few references to disability and access in the mainstream planning texts, and while Cullingworth and Nadin (1995), Bruton and Nicholson (1987), Hall (1984), Rydin (1993), and others, represent the established genre, none of them says anything about the coercive practices and ideologies of planning in reinforcing and (re) producing disablism in the built environment.

That such lacunae exist is not surprising given the much wider societal marginalization of disabled people, discussed in chapter 1. This lacuna has, historically, been reinforced by a range of influential planning theories and practices which have served to marginalize and/or ignore interactions between people with disabilities and the built environment. One such conception, for instance, relates to the emergence of new settlements. The garden city ideal is important in this respect and its originator, Ebenezar Howard (1898, 1902), conceived it in the aesthetic and moral values of industrial philanthropy of the time, the propagation of the romanticization of the 'rural', of the virtues and health-giving properties of fresh air, sunlight, open spaces, and the carefree lifestyles which (apparently) ensued.[3] Indeed, for Howard the garden city was a means 'to raise the standard of health and comfort of all true workers of whatever grade', a theme which underpinned the development of a whole range of model towns which sprang up in the nineteenth century to (supposedly) provide the social and environmental conditions conducive to (re) producing the true, or, 'able-bodied', worker.[4]

Indeed, Howard's (1898) conception of the garden city involved a particular construction of health, one which counterposed 'health' to 'disablement' and which, as Figure 6.1 illustrates, consigned the 'insane', the 'inebriates', and the 'epileptics', to segregated places outside of the main population centres. For Howard, the segregated 'solution' no more than reflected the dominant thinking and social practices of the time, yet his conception of disability was progressive in seeking to give people with disabilities equal rights to enjoy what he thought of as liberating and moral environments. Thus, as Howard describes it, 'dotted about the estate are seen various charitable and philanthropic institutions . . . in an open healthy district . . . and it is but just and right that the more helpless brethren should be able to enjoy the benefits of an experiment which is designed for humanity at large' (p. 20). While a paternalistic attitude underpins Howard's thinking, the spread of new settlement ideals became one of the lynchpins of twentieth century planning, yet, as a range of authors has noted, they were rarely sensitized, in the way Howard was, to the needs of specific groups. In certain respects, they became an important component in (re) producing disablist socio-economic structures, and the post war new towns were remarkable for their social homogeneity, for perpetuating social divisions and for seeking to provide the ('able-bodied') labour pools for the waves of productive decentralization from the larger metropolitan areas.

Throughout the 1940s and 1950s, plan making was implicated in (re) producing the modernist city, of separating functional spaces and, as Healey (1995) comments, 'producing environments functional to the needs of industry, providing for the material welfare, moral improvement, and aesthetic enjoyment of citizens' (p. 261). Yet this was the period of rampant technical proceduralism, a concern with what Healey terms a 'moral utilitarianism'. Indeed, as chapter 4 discussed, the propagation of modernist aesthetics and values in the built form was premised on a conception of a unified public interest and a homogeneous citizenry for which the planner 'as expert' could plan for (in the certainty of the predictable patterns of their clientele). Local plans were, therefore, seen as expressions of the (societal) totality, the embodiment of a techno-rational proceduralism apparently reflecting and (re) producing the core, central, values of society. The extent to which plans were sensitized to diversity, difference or to divergent value systems, then, was limited, while, as Healey and others have argued, a social conservatism supportive of development interests has been, and continues to be, the broad value-basis of the local plan.

While planning throughout the 1960s maintained an elitist distinction between the planner (as expert) and the public, the apparent expertism of local planning processes was called into question time and again by the recurrent disjunctures between stated objectives and outcomes. Economic crisis in the 1970s, for instance, undermined the ability of the planning system to maintain a control and command function while, increasingly, planners were forced to confront the social distributive consequences of their plans. Competing groups began to seek greater redress through the planning system, its value-neutrality was challenged, and the claimed status of the local plan, as somehow a technical arbiter between competing interests, was seen to be flawed. Greater participation in the system was called for, while radical critiques emerged to try and (re) direct the basis of the system away from privileging development interests towards a wider, more inclusive, social base. The claimed technical, apolitical, neutrality of previous periods, then, was threatened, yet local plans, even during this period, rarely reflected the ferment of debate. Participation in planning by

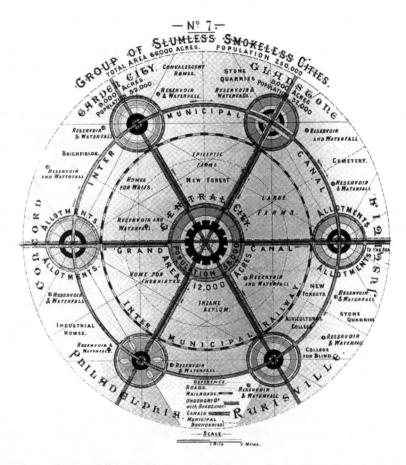

Figure 6.1   Ebenezer Howard's conception of the social city reflected the visions of a section of late nineteenth century society who were concerned with creating the 'good environment' for all citizens. Howard's ideas were based around the need to reform private land ownership and his involvement with the Land Nationalisation Society indicates how he conceived of using land and rent in engineering social change (see Ward, 1994, p. 24). Yet the scheme is underpinned by moralistic values, of particular forms of social order reinforced by spatial segregation. People with disabilities are acknowledged yet they are placed outside the cities away from the 'normal' population.

the public remained hierarchical, while the content of local plans never seemed to move beyond bland, general, statements of intent. Even where the land use implications of particular proposals were specified for particular social groups, like people with disabilities, there was always the insurmountable problem of not being able to resource some of the key policy intentions.

While local planning practice never attained significant powers, even during the 1970s, throughout the 1980s local planning departments experienced a progressive weakening of the local plan as a statutory and binding planning instrument. A range of directives increasingly placed the onus on less prescriptive policies and programmes while seeking to 'liberalize' local regulatory powers to enable, so it was argued, greater leverage for the private development industry. A number of strategic moments are illustrative of this more general trend which Thornley (1991), and others, have described as the emergence of an anti-democratic authoritarianism intent on asserting the values of the neo-liberal right (also see Rydin, 1993). In particular, a succession of governments has put emphasis on creating a 'streamlined' planning service better geared to the tasks of assisting economic recovery, in essence, utilizing a strategy of revitalizing the market as the central principle of social organization. In practice, this has been exemplified by the relaxation of the General Development Order and Use Classes Order, the transfer of planning powers to Urban Development Corporations, and the greater stress on market criteria in development control decisions.[5] In this sense, a market utilitarianism, as other chapters have intimated, sought to (re) establish the contours of the land use planning system.

Indeed, throughout the 1980s, government advice and circulars emphasized the 'legitimate' spheres of planning as ones of physical land use matters, somehow (re) directing the system towards the notion that technical, procedural concerns were of paramount importance.[6] Planners were exhorted to 'speed-up' the system, expedite decision making and create the conditions for investment and economic development. Public participation was squeezed into tighter time frames, it became more centralized and, as circular 23/81 argued, planners, in their preparation of local plans, 'need not necessarily consult and/or undertake publicity at the survey stage'. The period for decisions to be made on planning applications was reduced, while developers, far from sensitized to the needs of disabled people, were given greater powers to specify the form and content of development schemes. Planning for residential development, for instance, led the way with the emergence of joint (public-private) housing studies where house builders became increasingly involved in detailed site allocations at a local level.

Social distributive issues were also discounted by the DoE in the 1980s as 'irrelevant' to land use planning and, as circular 22/84 stated, 'it is inappropriate to include in structure or local plan policies any proposals which might lead to the assessment of planning applications on the basis of the identity, personal needs or characteristics of the individual' (DoE, 1984, p. 44). Planners were also directed not to produce policies favouring one group over another, yet such exhortations were contradictory in that government directives were simultaneously forcing planners to prioritize the requirements of the development industry. Thus the White Paper, 'Lifting the Burden' (DoE, 1985c), stated that development plans are useful because they 'can assist developers and the business community'. In addition, the status of plans was reduced to a form of 'developers' prerogative' in that, as Thornley (1991) notes, the development industry was more or less told by government that their market knowledge would be permitted to override the content of the development plan and/or any other form of supplementary planning guidance.

The Planning and Compensation Act (1991), which reasserted the primacy of the plan in the determination of planning applications, seemed to signal a retreat from the market utilitarianism of the 1980s. For Healey (1995), and others, this seems to indicate an apparent shift from market based criteria to a greater concern with environmental quality and community issues, and a re-emphasis on strategy and plans. In particular, Healey notes how a plethora of new strategic objectives – environmental, development, social and community – has led to increased debate about policy co-ordination, while there is now a greater air of expectation that planning can deliver people from many of the harmful externalities associated with modern economies. Indeed, the government's (apparently) renewed faith in planning as a tool of environmental management and change seems signalled in the DoE's revised 'Planning Policy Guidance 12: Development Plans and Regional Guidance', a document which encourages local planning authorities to consider issues like the 'revitalization of local economies', while, simultaneously, taking into account 'social distributive' issues.

Yet, for people with disabilities, and other marginal and oppressed groups, their status within the local planning system seems clear. For instance, in mid 1993, the Royal Town Planning Institute (RTPI, 1993) submitted a document to the junior minister of the Department of the Environment (DoE), David Curry, requesting that prescriptive policies, or policies which make binding, unequivocal, statements, appearing in Unitary Development Plans (UDPs) and/or local plans be given the department's support.[7] In particular, the RTPI was seeking to be allowed to go beyond the advice of Planning Policy Guidance 3 which states that local authorities can only 'seek to negotiate' accessible housing, towards the use of directives in plans which permit them to make access mandatory. However, Curry's (1993) response was clear in arguing that 'your document is wide-ranging and ambitious. It goes a great deal further than the revised planning guidance of PPG 3 and gives detailed guidance on a number of issues which in my view remain unresolved . . . for reasons I know you will understand, the Department would have to object to prescriptive policies appearing in local plans and UDP's.' In stating this, then, Curry was giving government support to the efficacy of market utility, for access only to be provided if the developer/applicant wishes to do so, and of the right of applicants to receive planning permission unless good reason to the contrary could be found. In short, the historic support of government for private property rights and the development industry was being reasserted.

LOCAL PLANNING PRACTICE AND ACCESS:
A CRITICAL ASSESSMENT

A paradox is that while the local planning system has been subjected to a series of decontrols and dictates seeking to reduce its influence on the content of social distributive issues, planning for disabled access seems to have become a more central element of the work being carried out by local planning authorities in recent times. Most local planning authorities provide some statements about access in their local plans, and there is a greater awareness of access issues than in the mid 1980s (ACE, 1994b; Imrie and Wells, 1993; Imrie, 1996c). Yet the challenges for local planners in practice in the 1990s remain large and daunting. Foremost is the continuing battle between local and central government, while the discretionary flexibility of the planning system remains a mixed blessing, enabling, on the one hand, a high level of local interpretation of statutes and, on the other hand, maintaining central direction and control. Moreover, whatever the government rhetoric about social, community and environmental values, there is little evidence that economic concerns and values are being overturned or, indeed, being fashioned to facilitate the attainment of a wider range of social and environmental goals such as accessible environments.

Yet, despite this, planning for access is part of the daily routine of local planning departments, although our understanding of the interrelationships between the components of the local planning system and planning to secure accessible environments is limited. In particular, there is scarce knowledge of the ways in which planning authorities use the local plan as a mechanism for seeking to secure access and whether or not access provisions can be attained through it. Moreover, in a context where developers have some prerogative to challenge planning conditions placed on a planning permission which departs, materially, from physical land use matters, it is unclear how far, and with what success, access is being secured through the utilization of conditions by development control officers. Indeed, there is some concern that many local planning authorities are abrogating responsibility for access by 'devolving it down the line' to building control departments, where, as previous research suggests, the issues tend to be treated as a matter of 'design configurations'. In amplifying such themes, the rest of the chapter considers the role of local plans, development control and building control in influencing the form and content of access for people with disabilities in the built environment. I base the discussion around secondary data and the interview and postal surveys outlined and discussed in chapter 5.

Local planning: an under-utilized system?

As a range of researchers has argued, many local planning authorities pay lip service to access issues, characterized by vague statements in local plans and an unwillingness to use planning conditions or other powers to enforce access provisions (Imrie and Wells, 1992; Millerick and Bate, 1991). Such research indicates that the local planning system has generally been ineffectual in either placing access issues on the agenda or developing positive programmes and policies to secure access provision (Imrie and Wells, 1992, 1993). Indeed, many planning authorities prefer to leave access issues to Building Control departments, arguing that they do not possess the requisite statutory powers to enforce access provision. Yet, in seeking to transform the interrelationships between local planning practices and planning for access, the Royal Town Planning Institute (RTPI, 1993) has declared that 'good access is a universally important town planning issue: good access is good planning' (p. 2). One of their Advice Notes (RTPI, 1988), for instance, points out the opportunities which local planning authorities have, especially in using local planning and development control powers to improve accessibility in the built environment, although its tone is reductionist in portraying the 'access issue' as, essentially, a design problem.

The local plan is increasingly being used to highlight access as a planning issue and as one of the interviewees commented, 'five years ago we had nothing on access in the plan, now it's seen as absolutely essential'.[8] Others concurred with this, and as an officer from a city authority argued, 'the local plan is an important instrument for us and access is an integral part of all of our planning policies . . . it wasn't like this a few years ago though'. Of the 22 interviewee authorities, 12 had developed access policies in the last six years. Such figures concur, in part, with the postal survey, and, when asked whether or not statements on access were included in their current local plans, 81 (67.5 per cent), of the postal survey sample replied that some form of statement existed, while 33 (27.5 per cent) said that they had nothing. However, 20 of the latter authorities did say that revisions to their local plans and/or their new UDP would contain some policy statements on access. Such figures contrast with the Welsh Council for the Disabled (WCD, 1987) survey of local planning practices in the late 1980s with 71 per cent of their sample having no reference to access issues in their local plans. Such contrasts, then, suggest that access is increasingly an integral component of local plans, and, as a large city authority noted in interview, 'the plan is only one of many instruments but it is important because it gives developers a clear signal of our intent . . . we see it as indispensable'.

Yet, as Table 6.1 indicates, the range and types of statements used in local plans vary among local authorities, both in terms of what they seek to incorporate into access and also the strength of the statements made (i.e. whether they are vague and general or prescriptive and precise). In particular, the majority of access policies lack a prescriptive style with statements tending to be bland and vague with few specifications concerning implementation.[9] Such authorities are what I term averse where phrases like 'will make every effort to facilitate access' or 'will seek to encourage access' or 'we will invite developers to provide access' are used (also see Table 6.4). Other authorities, in reinforcing the underlying voluntarism, utilize statements like 'the authority

Table 6.1   Local planning policy statements on access for people with disabilities

Type of local authority	Example of statements in local plan
AVERSE	
Operates with a biomedical model of disability, or disability as being derived from a physical and/or mental impairment	'The needs of people with physical and sensory handicaps will be considered in the undertaking of traffic management, highway construction and public transport projects'
Very few statements on access in the local plan	
Never uses planning conditions to secure access	'requiring cash dispensers to be accessible and usable by all residents, including people with disabilities, wherever practicable'
Major concern is to secure investment to support the local economy. Attitude is that insisting on access will scare away developers and much needed investment	
Typically a rural local authority and/or area with severe economic problems	
PROACTIVE	
Operates with a social model of disability or disability as a form of discrimination	'Proposals for development, including new buildings, or for the change of use, and alteration or extension of buildings . . . will be required to provide suitable access . . .'
Brings to the attention of developers all the statutory requirements on access	
Seeks to negotiate with developers and persuade them to give more than is required by statute	'Where development includes 12 or more dwellings, encouragement will be given for at least 20% of dwellings being provided to mobility standard'
Major concern is to secure development but not at the expense of neglecting statutory duties on matters like, for instance, access	
A mixture and range of localities	
COERCIVE	
Conceives of disability as an equal opportunities concern	'Provision of and facilities for people with disabilities is an important material consideration in assessing development proposals'
Insists on access provision	
Will seek to use all available planning instruments	'Planning Permissions for changes of use to premises covered by the provisions of the Disabled Persons Act will normally be conditioned to ensure that building works required to implement the permission make provision for access for the disabled'
Will not hesitate to use planning conditions relating to access	
Typically a left-wing, city, authority	
	'The Council will have regard to the needs of disabled persons and blind persons when placing lamp posts, bollards, traffic signs, apparatus or other permanent obstructions in the street'

Source: Interview Surveys, Postal Survey, Various Local Plans and UDPs.

will make developers aware of the relevant legislation' while some suggest that all access considerations 'should be referred to the building regulations and building controls department'. These, however, tend to be more proactive authorities in that they seek to use some of their planning tools in securing access in a way that the averse authorities do not. Other authorities, however, are bolder and, as one London borough stated in interview, 'appropriate access facilities for people with disabilities will be included in all developments which disabled people use'. Statements like this tend to reflect a coercive attitude, a political assertiveness on the part of authorities who use them. More often than not, such statements are the product of a particular local, left wing, politics seeking to align access to a wider social and economic agenda.

Such agendas tend to be found in the larger metropolitan authorities where the local plan is considered one instrument among many in pursuing access. Indeed, as one city authority noted, the function of the local plan statement is to make it clear to developers and architects that before they start any development they fully consider access for disabled people. As one authority noted, 'if issues are specified 'up-front', and at an early stage in the application procedure, then most developers will respond in a positive manner'. Such authorities were also aware of the significance of the 1991 Planning and Compensation Act in emphasizing the centrality of the local plan in determining planning applications, while, as one officer noted, 'supplementary guidance is part of our strategy . . . we use all sorts of design guides and pepper developers with all sorts of advice'. Yet, of the 22 interviewee authorities, only 4 have utilized some form of supplementary guidance. As Reeves (1995) suggests, the use of supplementary planning guidance, which is supposed to provide additional details about access policies in local plans, is not very evident and, of the 210 plans submitted to the RTPI in January 1995, only 14 contained supplementary guidance on access matters.

However, the majority of the interview sample, which were predominantly non metropolitan district councils, had reservations about using their local plan as a positive, proactive tool for promoting access issues in the planning process. While they could see little reason not to include some statements in the local plan on access, they were sceptical about making them 'stick' and, as one authority said, the 'acid test is making the provisions work in practice'. In this sense, some local authorities regard policy statements as 'going through the motions' rather than enabling them to pursue any serious intent. In part, such attitudes were clearly related to past experiences whereby the presumption in favour of developers, operating throughout the 1980s, was still assumed to be dominant today. As one of the respondents said, 'we've been pummelled by the government and we don't have much confidence in going against any central directives'. Moreover, many of the planning authorities had little faith that their policy statements would be acceptable to the development industry: as one commented, developers are out 'to make money and won't entertain many of the strictures placed on them by local plans'.

In particular, there seemed to be some confusion amongst planners of the viability of the local plan as an instrument for furthering access issues and, as one authority said, 'circular advice from government seems contradictory . . . we have difficulties making sense of it and working out what to do'. There was also little knowledge of practices elsewhere in the United Kingdom, involving local authorities successfully 'persuading developers' to accord with access provisions, and, certainly, a number of authorities I spoke to found the idea of writing access statements into their local plans as 'quite novel' yet 'probably unworkable'. While education and training have a role to play in demonstrating the possibilities of the local plan as a statutory instrument, most of the sample feared that onerous restrictions imposed by an authority could lead developers to go elsewhere or simply to abandon a proposal. As one authority noted, 'we get so little investment here anyway that imposing access restrictions isn't really on'. Others concurred with this, especially noting that in a recession 'any additional costs to an applicant can kill development stone-dead'. As one authority noted, 'we've got little confidence in the DoE and anything we try and do to make developers incorporate access will be wiped out in an appeal . . . so we try and negotiate . . . it's much better this way'.

The claimed non viability of the local plan was also a concern of those who felt that the DoE was being over-zealous in policing the content of their plans. Much of the evidence from the submission of the UDP's to the Secretary of State seems to suggest that prescriptive statements concerning access are being deleted and/or re-worded to change (and water down) their emphasis. Evidence from the last chapter indicates a reactionary stance being taken by the DoE and as an access officer in a London borough recounted to me, 'we placed a range of directives about access in our UDP but they've all been taken out and we've been asked to replace them with a bland statement . . . how can that help . . . most of us just feel fed up with it'. As a range of respondents implied, the content of the UDPs is being tightly controlled by the DoE and little or nothing which would challenge the voluntarism of existing access regulations is being conceded. As one authority commented, in relation to their housing allocations policy in their UDP, 'we've requested certain things and we originally insisted that we 'must see' allocations set aside for people with disabilities . . . but the minister didn't like this and now its been changed to 'would like to see'. This, then, is suggestive of a circumscribed, bounded, system where localized discretion is tightly constrained by central directives.

The possibilities for development control and access

Throughout the 1980s, the importance of market utility in planning practice was re-emphasized time and again and its centrepiece, circular 22/80, signalled the government's intent of speeding up development control decision-making while making it more responsive to developers (Rydin, 1993; Thornley, 1991). In the context of planning for access the circular suggested that the planning process had to devolve 'responsible decision-making', about the design and external configuration of buildings, to developers or applicants. As Thornley (1991) notes, the underlying rationale of the circular was to stop what the Conservative government of the day regarded as interference by planning bureaucrats in decisions which the Secretary of State for the Environment at that time, Michael Heseltine, felt were best left to the market. For Thornley, the purpose being pursued was one whereby 'the fact that the developer will have to sell the development on the market will be the control over acceptable design and will also allow greater innovation, not to mention reduction of delay' (p. 148). In this sense, development control was being directed towards a facilitative role or what Thornley has characterized as 'direction by the market'.

In particular, some of the strategic exclusions of specific developments from access regulations and/or development control are indicative of the wider ethos of maintaining forms of market utility as the guiding force of land use developments. As chapter 5 indicated, domestic dwellings are excluded from the requirement to conform to access criteria while only new non domestic buildings are required to conform with Part M of the Building Regulations. Moreover, while, historically, development control practitioners have tended to claim a value-neutrality, in that their techniques and procedures are a guarantee of an (objective) impartiality, others have pointed towards the partiality of the 'decision rules' which promote some interests while disadvantaging others.[10] Thus, as documented in the last chapter, certain public sector interests are protected, and excluded, from access regulations without any clear reason why this should be so, while central government has cultivated strong reserve powers to impose its own criteria and judgements. In addition, central government, through their circular advice, have the ability to tightly police development control decisions.

Not surprisingly, such socio-political conditions have influenced development control practices on access. Thus, for the majority of local authorities which were interviewed, there was a reluctance, for instance, to use planning conditions to force developers to take account of disabled persons' needs. The general view was that such a planning condition would be unworkable and that any appeal brought against an authority would be won by the developer.[11] One authority commented, 'most developers would just ignore the condition anyway and go ahead with the development regardless'. Respondents from the postal survey were also sceptical about their ability to use conditions. For instance, as Table 6.2 indicates, 37 (30.8 per cent) of the respondents said that they have never placed a planning condition on an application or, as one authority commented, 'we'd never get it past the DoE . . . they don't like us doing this sort of thing'. Another authority concurred with this in noting how 'Planning Policy Guidance 1 (PPG 1), issued by the DoE, indicates that conditions are not relevant so what's the point', while another authority, seemingly dismissive of planning conditions in any form, at any time, remarked, 'why would it be felt necessary to include planning conditions when the DoE have placed access into building regulations'.

Table 6.2   The proportion of planning applications that have planning conditions attached to them concerning access for the disabled

Proportions of planning applications with conditions on access attached	Local Authorities Numbers (%)
All of them	0 (0.0)
50% to 99% of them	1 (0.8)
25% to 49% of them	1 (0.8)
11% to 24% of them	7 (5.8)
Less than 10%	48 (40.0)
Never use conditions relating to access	37 (30.8)
No Response*	26 (21.6)

*Many said that they could not quantify their response and had no knowledge to answer this question. It seems that few, easily retrievable, records are kept by local authorities recording this type of information

Source: Postal Survey, 1994-95.

Yet, this comment really reveals a remarkable ignorance of the possibilities of planning conditions while the remark relating to PPG1 seems inappropriate given circular 10/82's advice that access can 'be material'. In particular, the comments from the sample authorities serve to highlight variability in practice and also much confusion about what is possible. Moreover, different authorities seem to be working with quite contrasting interpretations of what is permissible given the existing legal structures. Indeed, many of these who did not feel the need to use conditions referred to, time and again, the efficacy of the building regulations and were obviously placing great faith in their ability to ensure access. In part, such attitudes reflect the tightly defined way in which the DoE conceives of when and how planning conditions should be used. In particular, ever since the issuing of circular 1/85, 'The Use of Conditions in Planning Permissions', local planners have been increasingly circumspect about using them. This relates to the provisions of the circular which states that 'conditions should not be used unnecessarily; a condition ought not to be imposed unless there is a definite need for it and planning permission would have to be refused if that condition were not to be imposed' (para 12).

There was evidence, however, of a number of authorities using planning conditions which cited access as one of the material concerns. For instance, one authority regularly places a condition which states that 'the plan should comply with the access officer's requirements'. Indeed, this particular authority was involved in an appeal at the time of interview over a case where a shop owner had refused to remove a step into the premises. Another authority, as a matter of course, regularly places a planning condition which asks for the applicant to ensure access for the disabled. Yet, as they admitted, it is usually left to the discretion of the applicant about what to provide, with the authority giving little design guidance or direction. In particular, economic circumstances seem to dictate, in part, the extent to which planning conditions are used, and there is a real division between localities with high and low levels of demand for development. The former have more latitude to refuse applications which do not accord with planning conditions, with the realistic expectation that another application 'will come along'. In contrast, the latter, in seeking to attract limited mobile investment, often feel unable to set conditions which might scare developers away. In this sense, levels of market demand for development clearly influence the imposition of planning conditions. In the present voluntaristic climate, many planning authorities will remain reluctant to press conditions which might direct investment elsewhere.

Moreover, refusals of planning applications solely on the grounds of access were more or less non existent in the sample authorities. Where refusals have occurred, access has only been one of a number of reasons for the decision and usually of subsidiary significance. For instance, only 15 (11 per cent) of the postal survey authorities had, within the previous five years, refused planning permission on an application citing access as one of the criteria for the decision. Where permission had been refused with access cited as a reason, the dominant reasons for refusal tended to range from 'the development detracts from the character which the conservation area seeks to retain' to 'the amenity of the area will be adversely affected'. Access was subsumed under such stated conditions and, presumably, on its own would have been difficult to enforce. As a postal survey respondent said, 'we've had problems before trying to make access stick with conditions . . . so we don't really use them . . . negotiating is best'. In such instances, then, it seems that the local authorities did not have the confidence to refuse permission solely on the grounds of access and, as one authority commented, 'such a decision would be laughed through the appeal process and we'd never make access objectives stick . . . it isn't taken that seriously'.

This research concurs with the Welsh Council for the Disabled (WCD, 1987) survey in that the refusal of a planning consent on access grounds tends to be regarded as illegitimate by most planning authorities. While the WCD survey showed that no local authority had refused a permission on access grounds, one of my face-to-face sample authorities had refused permission partly on the grounds of an unwillingness to comply with access conditions. On a number of occasions this authority has refused planning permission for new shopfront developments, although access was always one amongst a number of reasons for refusal (and, like the postal survey authorities, was never the primary one). Moreover, while authorities are reluctant to refuse planning permission solely on access grounds, two of the sample authorities had taken action to force developers to comply with the original permission. For instance, a recent office development in one of the local authorities provided splendid curving ramps for wheelchairs, yet, in the original finish, placed six inch steps up to the entrances of the building, hence defeating the purpose of the ramped sections!

However, it could be argued that resorting to planning conditions to secure access marks a failure of other, proactive, parts of the planning system to achieve the same ends. As one authority commented, 'we prefer to be more involved at the pre-planning application stage where we have more power to persuade developers . . . they're usually sympathetic'. For this authority, like many others, the utilization of a condition represented a failure of negotiation and, by implication, a failure of them and the planning process to build-in, up-front, access issues. A wide range of approaches was evident from interviews. One authority, for instance, uses a design guide produced by the Access Committee for England, and one specific application for the conversion of a warehouse into shops had a condition attached specifying the criteria to be used in developing access for disabled persons. However, as the authority pointed out, 'it was left to the discretion of the developer about what to provide and they ended up only providing for wheelchair users'. One authority had used a Section 106 agreement in securing access provision, requiring the developer to place reserved car parking spaces for disabled persons in a private car parking development, thus illustrating the possibilities of using this part of the Town and Country Planning Act in securing some access provision. Others variously provide a full range of access guidance and notes at the preliminary stages of the planning application and, as an authority commented, 'developers can be more responsive than many seem to think'.

The limitations of the building regulations

The major legislative mechanism for securing access, as the previous chapter intimated, is Part M of the Building Regulations in England and Wales (or Part T of the building standard regulations in Scotland). Part M, introduced in 1987 and extended in 1992, seemed to be a radical breakthrough in access legislation in that it widened the scope of control to all public and commercial buildings. Yet Part M is a weak and reactive piece of legislation and, as some observers have noted, it couches regulations in a vague and ambiguous manner which does little to clearly define what is possible (Barnes, 1991). Indeed, one of the more problematical aspects of the 1987 regulations was the exclusion of visually impaired and/or blind people, and the hard-of-hearing and/or deaf, from consideration while only the 'principal entrance storey' of buildings was required to be made accessible. While such exclusions were abolished in the revision to Part M in 1992, partiality and omissions are still the defining features of the building regulations, seemingly reinforcing the government's concern to minimize cost to the private sector.[12]

As Table 6.3 indicates, 47 (38 per cent) of the postal survey respondents commented that they found the building regulations 'helpful' in providing them with a relevant framework for securing a more accessible environment. As one of the respondents noted, 'they provide a neat and tidy way of responding to access . . . it's all laid out with rules and regulations to follow'. For others, the regulations provide 'the bottom line' while the general attitude was one of 'any legislation that makes inroads into the improvement of access has got to be helpful'. In this sense, the building regulations tend to be viewed as limited and

Table 6.3 The helpfulness of the building regulations in providing a framework for securing a more accessible environment.

Helpfulness of building regulations by category	Local Authority responses Number (%)
Extremely Helpful	36 (30.3)
Very Helpful	28 (23.5)
Helpful	47 (39.5)
Unhelpful	3 (2.5)
Very Unhelpful	0 (0.0)
Extremely Unhelpful	2 (1.7)

Note: There were 4 non respondents to this question.

Source: Postal Survey 1994-95.

limiting, yet, as one respondent suggested, 'better than nothing', an attitude which is partially explained by the relative absence of any real form of control concerning access prior to 1987. However, some pointed to contradictions and difficulties. As one postal survey respondent said, 'the standards are sometimes in conflict with new equipment, i.e. door widths' while another commented that 'they give us little enforceable power, we need the all inclusive American approach . . . all premises to be accessible to disabled people'.

The majority of local authorities, both postal and interview, concurred that the building regulations are their main instrument for securing access provision. Yet there was widespread divergence in its utilization. As one authority noted, 'we use it about a third of the time, but, of this, less than 25 per cent of the applicants will actually conform with what we want'. However, most respondents argued that it was difficult to enforce the provisions and I gained direct evidence only of the city authorities taking enforcement action to gain compliance by developers. As one respondent added, 'enforcement is a tricky job, we really rely on the vigilance of the access group to report that a building hasn't come up to scratch'. Another authority agreed with this noting that while the plans are correct, 'on-site everything is not as we wanted'. Indeed, this authority, while committed to enforcing Part M, often found applicants resistant to regulatory control. For instance, one applicant, a solicitor, decided to give the authority a run for their money in court on Part M, after defying a condition to provide various access provisions in a new development. The authority, in attempting to put pressure on the solicitor, took councillors out to the site to show why a summons had been brought, indicating stepped entrances around the building with the disabled toilet on the second floor with no lift access! To reinforce the seriousness with which the authority viewed this transgression, a visit was made by the local access group and pressure was placed on the solicitor to conform to Part M. As the authority commented, 'word has now gone around to developers that the council is serious about clamping down on non-compliance with Part M'.

However, 7 out of the 22 interview authorities admitted that they had taken little or no enforcement actions against known transgressors of Part M. The main difficulty for developers, as many of the respondents said, is one of cost or, as one authority put it, 'developers will always argue against Part M on cost grounds, even where those costs are really only marginal'. As another authority concluded, 'most developers can't see any demand for access features . . . although more of the shops and retailers are now leading the way'. Indeed, this is where the hope lies in overcoming the voluntaristic codes laid down in Part M, that developers will increasingly see disabled people as an important client group who should be served in the same way as the 'able-bodied'. The postal survey authorities generally had a positive attitude towards developers' responsiveness to access and Part M, claiming that most will accede to access directives after a period of negotiation. In particular, the widely held assumption about developers, that they seek to maximize their profits and will do everything to resist the high and prohibitive costs in creating accessible places, seems, in part, misleading. Certainly, the government presents access as a potential cost burden, while specific lobby groups, like the House Builders Federation, campaign against it (see, House Builders Federation, 1995). However, as one of the postal survey respondents noted, 'there is no such thing as a typical developer . . . they're all very different and the attitude of a small property company is often at odds with that of a large house builder . . . many are quite happy with Part M'.

This reflects a common situation, indicated by a range of respondents, whereby there is a high level of reliance of the planning system on Part M to regulate and control for access provisions. As one of the survey respondents argued, 'Part M is the only really enforceable legislation with clearly set out provisions. However, it does not go far enough and is often used as the excuse not to deal with access at the planning stage which I think is essential'. Others concurred and, as an access officer noted, the building regulations are 'very unhelpful in terms of existing uses . . . and the current DoE insistence that they are the only appropriate means for ensuring access is extremely unhelpful in enabling us to specify anything on access when existing buildings are adapted . . . this seems a nonsense'. In particular, the access officer for one London borough was adamant that the regulations were less than helpful. He commented, 'when I was director of the Access Committee for England it was clear that there were two big issues we had to confront, namely how to make housing accessible and how to bring in access requirements for existing buildings to be adapted . . . everyone knows that existing buildings are the real problem but the building regulations hardly addresses this . . . so, they're not much good really.'

CONCLUSIONS

Throughout the 1980s, and early 1990s, local planning authorities began to incorporate access policies into their local plans while seeking to place access issues more firmly into the criteria used for decision making in development control. While, in part, this reflected the exhortations of the RTPI, it was also related to heightened pressures being exerted by people with disabilities to gain some participatory forms of representation within the power structures of local authorities (see chapter 7). Yet, despite a period of change, the majority of authorities seem to have consistently understated their powers to secure access for the disabled, while many seem reluctant to use either their local plan, or planning conditions, as tools for promoting access issues. Similar reticence was also identified in the mid 1980s (Welsh Council for the Disabled, 1987). Indeed, far from using their limited powers, planning officers were, in the main, continuing to utilize Part M of the Building Regulations to secure access provision, despite the limitations of the underlying legislation. Yet my research detected dissatisfaction amongst planning officers who, in using Building Control as the main mechanism for access decisions, felt that an application 'had gone too far by then' to significantly change its design and form for the benefit of disabled users.

While evidence of this type suggests a uniformity of approach by local authorities towards access, there are significant variations in practices and policies and the evidence, both from this and the previous chapter, suggests that there are at least three different types of approaches by local planning authorities towards access and disability. As Table 6.4 outlines, the first is where access is of little or no concern and barely features as an issue let alone as a component of daily practice. Authorities which fall into this category are increasingly in a minority primarily because of Part M and the exhortations of the RTPI. Access is treated as peripheral and marginal, possibly a threat to investment with the potential to scare away development. They have no design guidance on access, and never raise access issues with developers. Such authorities seem all too ready to accept the liberal rhetoric that 'access costs', and are also typified by 'political apathy' towards access with local councillors showing little or no interest. As one interviewee stated, 'to my knowledge access has never been discussed in council meetings . . . no one here's interested in it'. Moreover, access groups do not exist or if they do they are weak, poorly resourced, and tend to operate with a charitable ethos (see chapter 7).

In distinction, a second group of authorities has introduced a range of piecemeal measures in response to access issues. Such authorities represent the growing majority and their approaches to access tend to be underpinned by the support of their councils, the designation of an officer to undertake access duties, often on a part-time basis, with the probability that some form of access budget has been created. However, such authorities tend to operate within the legislative boundaries of Part M, even celebrating, as one authority put it, 'its liberating potential', while reinforcing stereotypical and reductionist conceptions of disability by asserting the possibilities of design solutions in overturning problems of disabled access. As one authority commented, 'we've done a lot to help the disabled by removing steps and insisting that the technical standards are met . . . that's as much as we can do'. Such authorities,

Table 6.4   A typology of approaches to access for people with disabilities in local planning authorities

Type of local planning authority	Characteristics of approach to access
AVERSE	Access potentially threatens inward investment and economic development
	Access is a minority issue and affects only a small proportion of the population
	Wholly reliant on Part M to guide access decisions
	Local political system unaware of access issues and provides little support or encouragement
	Access groups either non existent or poorly organized
	No access officer, or where they do exist, usually performed part-time by someone in Building Controls or Local Plans
	No budget to support access projects
PROACTIVE	Access is seen as an issue to be considered because of the directives of government and the Royal Town Planning Institute
	Access is seen as one amongst many competing demands on officer time
	Appointed access officer, usually on a part-time basis
	Some small funds for access issues
	Reliant on Part M but will try to negotiate and bargain with developers for what are presented as 'access concessions'
	Some awareness by local council of access issues but remains peripheral and rarely discussed by local politicians
	Access groups usually exist but are often weak and poorly resourced
COERCIVE	Access is a right of all people
	Societal oppression and domination of people with disabilities is at the root of their marginalization within the built environment
	Providing good access benefits all not just a minority group
	The local economy will benefit by providing access. People with disabilities are consumers too
	Directive Planning Guidance in the Local Plan
	Design guidance on access with details appended to planning applications
	Access negotiated prior to the submission of a planning application
	Regular meetings between planners and local access groups
	Full time access officer, well networked within and between departments
	Active support from local politicians with key councillors
	Active access group (s)

Source: Interview Surveys and Postal Survey.

then, tend to conceive of a 'designer approach' as the most significant way of overturning inaccessible environments, while the general philosophy is that good management combined with judicious use of the existing legislation will achieve some gains for people with disabilities. In this sense, the efficacy of a (bureaucratic) technical-proceduralism is the dominant form and style.

In contrast, a small number of local authorities has rejected medical theorizations of disability. Instead, the socio-political nature of disability is asserted, with an emphasis on social and political campaigning as the main method of redressing the iniquities which confront disabled people. As one authority noted, 'we see disability as a social construct, with disabled people being oppressed by their surroundings . . . policies should address this'. Moreover, the local plan of one local authority states that the borough council's approach to disability is one whereby 'it is often useful to think not of disabled people but disabling environments or places which oppress'. Invariably, such authorities are aligned to a left wing political base, and tend to be concentrated in places like the inner London boroughs and the larger cities in the UK. In helping to achieve particular political objectives, access budgets tend to exist, while, invariably, policies are spearheaded by a full time access officer. Planning conditions will be used as a matter of routine, while precise, often prescriptive, policy statements on access will be placed in local plans. A grassroots disability politics, in the form of campaigning access groups, is also an important component of libertarian authorities while people with disabilities are usually well represented in some of the powerful, and strategic, committees.

Variation, then, is a feature of local authority access policies and practices, yet there is a sense in which access is still conceived of as a marginal function, often an irritant and a diversion of officer time from some of the strategic functions of the authority. In part, this is because of the ways in which planners feel their hands are tied by central government directives in seeking to negotiate access. In turn, planning officers tend to defer to developers and even where Part M regulations are attached to a planning application for a new development, there is often little follow up to see whether or not the developer has adhered to the access regulations. Most authorities concurred that enforcement was a difficult matter and, as one chief planning officer noted, 'we depend on the general public or the local access group to tell us about transgressions . . . we haven't got the staff or resources to do much enforcement work'. For the future, it seems as though the securing of access through the planning system will be a slow process, inevitably piecemeal, and geographically uneven.

NOTES

1. Coercive values, or what Young (1990) refers to as the hegemonic value systems of the core culture, of sexism, racism and disablism, are, therefore, evident in the specific policy tools, practices and outputs of local planning systems, that is, the technocratic

ideology. Thomas (1995), for instance, rationalizes the racist, what he terms 'colour blind', nature of the planning system in terms of the perpetuation and centrality of a bureaucratic ethic or the value-neutrality which many planners claim for their procedures and practices. As discussed in earlier chapters, an important component in depoliticizing social policy practices is the reduction of seemingly value-laden decision making (and its outcomes) to formal bureaucratic procedures, to a set of supposedly agreed rules which treat all in a 'fair and equitable way' (see chapters 2 and 3). As Thomas has argued, the centrality of a bureaucratic, procedural, ideology in planning is, in part, one of the reasons that many planners in practice resist the idea of positive discrimination for ethnic groups or other similarly disadvantaged minorities. Thus, as Thomas suggests, 'the initial widespread reaction is that the scrupulous bureaucratic formality which planners have perfected over decades is a guarantee that there will be no racial, or other, discrimination exercised within the planning system' (p. 135).

2. I am assuming that the reader is familiar with terms like 'local plan' and 'development control' and has some knowledge of the ways in which such systems operate in British local planning authorities. Instructive accounts of the British planning system can be found in Cullingworth and Nadin (1995) and Rydin (1993).

3. I would like to thank Huw Thomas for directing me towards this series of ideas and themes.

4. As Rydin (1993) has documented, Chadwick's report on the sanitary conditions of the working poor, and of the interrelationships between poor housing conditions and ill health, alarmed industrialists and a range of individuals attempted to improve the conditions of the workforce by building model towns. Thus, places like Port Sunlight and Bournville were developed and promoted as providing clean and healthy living places fit and able to reproduce the ablebodied worker.

5. The General Development Order (GDO) is a planning tool which itemizes whether or not a proposed development by a developer and/or householder requires a planning permission. Small scale developments will often fall outside of the GDO and the developer can proceed without applying for a permission. Similarly, the Use Classes Order (UCO) specifies classes of land uses. If a proposed development involves a change of use from one class to another class then a permission must be sought. Otherwise, if the change of use is within a class, no permission is required. Since 1979, successive governments have relaxed the strictures of the GDO and the UCO so making it easier for developers to proceed with developments without having to apply for a planning permission (see Cullingworth and Nadin, 1995).

6. In this respect, Planning Policy Guidance (PPGs) notes are an important part of government advice and they represent guidance by central government to local authorities on specified planning topics. They are advisory, not statutory.

7. The Unitary Development Plan (UDP) is replacing the old two tier structure, developed under the 1968 Town and Country Planning Act, of structure plans and local plans. The UDPs will have two parts. Part 1 is general policies, while part 2 sets out detailed proposals and guidance (see Rydin, 1993).

8. For instance, there are a number of well publicized cases which illustrate the possibilities for local planning authorities in securing access by utilizing a range of different planning instruments. One such case was the development of the Island Site in Hammersmith which involved the borough and developer using a planning agreement resulting in voice boxes being placed in lifts, all shops having individual access, with offices being wholly accessible to the disabled. Leach (1989) relates similar, progressive, local authority stances in Manchester and Lambeth, with the former characterized by

innovative dial-a-ride bus services, state of the art access features in the city centre, with active participation by disabled people in the formation of policy.

Another case, in Ealing, west London, highlights the possibilities of a plan-led system. In this instance, a replacement shopfront was developed without planning permission. Ealing Borough Council asked for a retrospective application from the occupier with officers indicating that a grant of planning permission would be likely. This was provisional on proper arrangements being made for disabled people, including a ramp rather than a step to the door. However, the application offered no change to the step and planning permission was refused. On appeal, the Inspector noted that the Council's planning policies on access for the disabled were the major issue. She concluded that, 'a ramp could have been provided in accordance with the Council's policies and that its omission adversely affects the rights of the disabled to unrestricted access to the premises'. Consequently, the appeal was dismissed, illustrating the potential power of the local plan as a binding statutory instrument.

9. A prescriptive statement in planning practice is one which signals a definite and binding intent. It uses language like 'must adhere to' rather than the more loosely phrased 'will be asked to adhere to'. Prescriptive statements try to tie developers down to a course of action.

10. Indeed, the particular discourses of development control are wrapped into a technical formalism which is difficult for those 'outside' the system to penetrate. In the context of the development control, the system, for instance, can barely be operationalized without some knowledge of planning law. Healey et al (1988), refer to the 'customs and practices of considerations and roles in the development control system' and castigate them for their complexity and of the vagaries of the administrative traditions and professional myths which, so they argue, 'privilege the knowledgeable'. This, in turn, echoes one of the key dimensions of societal oppression which marginal peoples have to confront in late modern societies, their estrangement from the knowledge bases which, in part, oversees aspects of their daily lives.

11. This seems to be a view shared by local authorities in Wales in the 1980s. For instance, the survey by the Welsh Council for the Disabled (1987) indicated that while 62% of their sample issued standard letters informing developers of their requirement to provide access for disabled people, the majority of local planning authorities (71%) were unwilling to use planning conditions to secure access, despite the message contained in Welsh Office Circular 21/82 (which encouraged Welsh local authorities to use planning conditions to enforce access regulations).

12. At the time of writing there are proposals to extend Part M of the 1991 Building Regulations. The main proposal is that reasonable provisions should be incorporated into new dwellings to allow future occupiers to invite people with disabilities into their home, or for occupiers to remain in their home over a longer timescale. The review is considering entries, doors, internal circulation, toilets, switches, and lifts. However, the clause of 'reasonable provision' makes it likely that little will change.

FURTHER READING

There is little or nothing written on access and planning for people with disabilities. Of the texts that exist, Imrie and Wells (1993) and Imrie (1996c) consider the interrelationships between planning for access and disability people in the built environment. Other interesting reads include Golledge (1993), Vujakovic and Mathews (1994) and Hahn (1986). On planning systems more generally, Thornley's (1991) instructive account of the restructuring of planning under neo-conservative regimes is worth looking at, while Rydin (1993) provides a general introduction to the mechanics of the British planning system.

7

'We're not the Same as You': Diversity, Difference, and the Politics of Access

INTRODUCTION

In the act of reclaiming the identity the dominant culture has taught them to despise, and affirming it as an identity to celebrate, the oppressed remove double consciousness. I am just what they say I am – a Jew boy, a coloured girl, a fag, a dyke, or a hag – and proud of it. No longer does one have the impossible project of trying to become something one is not under circumstances where the very trying reminds one of who one is. This politics asserts that oppressed groups have distinct cultures, experiences, and perspectives on social life with humanly positive meaning, some of which may even be superior to the culture and perspectives of mainstream society. The rejection and devaluation of one's culture and perspective should not be a condition of full participation in social life.

Young, 1990, p. 166.

Over the last twenty years or so people with disabilities have been struggling to overcome the dominant, ableist, values and practices of society (Barnes and Oliver, 1995; Oliver, 1990; Shakespeare, 1993, 1994). Such struggles have had to contend with the powerful and persistent nature of the paternalistic structures of the welfare state while seeking to overturn the dominant socio-cultural conceptions of disability, that is, that somehow disabling states are a function of physiological impairments and the (deviant) pathology of the individual. 'Naturalistic' explanations of disability have also been reinforced by the dominance of the charities, the public telethons, and the rehabilitation agencies, all of which have been and still are the dominant institutional forces in organizing the daily experiences of people with disabilities. In particular, the avowed apolitical status of such organizations has continually frustrated the efforts of disabled people to challenge and overturn the ableist values and practices of state and society more generally, while projecting and cultivating pejorative images of disability as a 'pitiful' and 'tragic' state.

In turn, welfare organizations have, as chapter 3 intimated, attempted to respond to the estrangement of disabled people from society by pursuing policies of integration and assimilation, a 'mainstreamism' which, ironically, has sought to deny difference and to subsume disabled people into the wider (popular) culture which, historically, has devalued them. For many disabled people, the resultant policies, while holding up the chalice of formal equality of opportunity, treatment and respect, have tended to reinforce the oppressed status of people with disabilities because the socio-institutional structures implicated in the (re) production of ableist values and practices have never been seriously challenged let alone dismantled. Thus, while employers in the UK, for instance, are legally obliged to 'integrate' disabled people into the workplace, by setting aside a jobs quota of 3 per cent of their workforce, few do so and even fewer are prosecuted and compelled to comply (Barnes, 1991; Barnes and Oliver, 1995; Morris, 1993).[1] Yet, the definitional basis of integration is problematical because, while it seeks to address, albeit inadequately, the marginalization of disabled people from participation in the formal (ablebodied) labour market, it wholly fails to tackle their marginalization and lowly positions within the workplace and/or the prejudices and practices which are implicated in the status of people with disabilities at work (see Barnes, 1991).

The false promises of society, of its integrationist ethos, have also served as a powerful instigator of an emergent 'politics of difference' amongst people with disabilities. In recent times, strands of disability politics have developed radical, self assertive, agendas seeking to project positive (self) identities and focusing attention on the structural causes of disabled people's oppression (Barnes and Oliver, 1995; Shakespeare, 1994). In particular, disability groups in the UK have begun to reflect and reproduce some of the more strident aspects of disability politics which underpinned the emergence of the Americans with Disabilities Act in the USA, a politics of protest based around the conception that only direct action would bring about the required changes in statute and wider socio-legal structures. National co-ordinating organizations have emerged, like the Campaign Against Patronage and Campaign for Accessible Transport, while direct action has proliferated with regular blockades of telethons and inaccessible transport. Such actions, as Shakespeare (1993) notes, are overtly political, 'showing that disability is a matter of social relations, not medical conditions' (p. 251).

However, the reality for many people with disabilities in the UK, and elsewhere, is their continuing political marginalization, their lack of representation in the more powerful institutions, their estrangement from democratic procedures, and the persistence of assimilationist and/or normalization policies by the state which are implicated in maintaining the hierarchical power structures of the welfare services.[2] Indeed, people with disabilities are denied basic democratic rights and, as Barnes (1991) shows, many disabled people in long stay hospitals are only allowed to vote if they satisfy particular conditions, one of which is the filling in of a declaration form which only gains validity if signed by a member of staff. As Ward (1994) suggests, this makes a disabled person's democratic rights dependent on the attitudes and practices of care professionals, something which people outside such institutions do not have to face. Moreover, while more disabled people are collectively organizing to oppose and contest the paternalism and authoritarianism of the state, such groups remain fragmented, poorly funded and generally ineffectual (Barnes, 1991; Imrie, 1996b). As Shakespeare (1994) and others have noted, despite the heightened levels of politicization of many disabled people, little has changed and the persistence of the culturally imperialist values of an ableist society remain firmly intact.

Yet, while the structures, for example, of the built environment are constitutive of the dominant (culturally imperialist) social relations, it is clear that such social relations do not remain uncontested by disabled people. In particular, it would be problematical to ignore the multiple ways in which oppressed groups contest both the production and reproduction of the built environment and, in doing so, contribute to its transformation. Indeed, people with disabilities are not merely passive recipients of the built environment, but actively seek to challenge and to change it (see Laws, 1994b). In this sense, an important theme in the study of disability and access is how disabled people are able (or not) to influence policies and processes which make a difference to their experiences of the built environment. In particular, in a situation where political power is exclusive and concentrated beyond the reach of most people with disabilities, and where institutional closures are seemingly resistant to responding to their needs, how can disabled people challenge the socio-institutional and political discourses and practices of (institutionalized) disablism. In considering such themes in the context of access, I divide the chapter into three parts.

First, I provide a brief overview and critical assessment of some of the under-lying factors which constrain the effective politicization of people with disabil-ities while providing a summary of the ways in which the state has attempted to institutionalize the growing demands of disabled people for accessible environ-ments. Second, I use case study material from two London local authorities to explore the contrasting ways in which the institutionalization of disabled people's demands for access has occurred and with what effects for the propa-gation of access issues. In doing so, a third section compares and contrasts the interactions between the local authorities and local disabled access groups, while assessing some of the organizational and practical barriers inhibiting the repre-sentation of disabled people's views on access.

INSTITUTIONALIZING THE POLITICS OF ACCESS AND DISABILITY: SOME PRELIMINARY OBSERVATIONS

As Shakespeare (1993) has argued, most of the emergent social movements, post 1960, were primarily concerned with resource allocation, economic exploitation and poverty, or with issues connected to the distribution of the economic surplus. The early movements in the USA, for instance, like gay rights and black activism, were assimilationist in seeking to achieve some form of recognition of marginal groups as being 'no different' from the rest of society, while their main target of campaigning was directed at overturning institutional discrimination. Similarly, feminism, in the earlier years, espoused a redistributive logic, a concern with equal pay, gender discrimination in the workplace, and campaigns over the absence, for instance, of workplace childcare facilities (see Matrix, 1984; Stanley and Wise, 1993). Likewise, the main thrust of the UK disability movement has been 'for more resources to be channelled towards disabled people and challenge the distributive logic of capitalism' (Shakespeare, 1993, p. 259). In this sense, the initial political focus of disabled and other groups was to attack their marginal status in the labour market, to claim parity with the rest of society, to, in effect, accede to assimilation into the dominant culture.[3]

Over the years, however, people with disabilities have widened the political agenda and while labour market and employment issues still form the focal point of campaigning, mobility, access and transport issues have become an important focus for direct action. Indeed, throughout the 1980s, the disaffection of disabled people in the UK, from the traditional purveyors and/or defenders of their interests, the charities, was increasingly expressed by wheelchair users, for instance, chaining themselves to buses, blockading streets and thoroughfares, while linking their oppression to their inability to gain access to the services, like trains and buses, that most people take for granted. For them, an 'identity' politics became an assertion of difference and of their rights to be treated according to the needs that their differences demanded.[3] Yet, while the politicization of many disabled people has been a remarkable aspect of recent times, concessions, as Barnes (1991) and others have noted, have been few and far between. As chapters 5 and 6 have indicated, this is largely because of the market-driven voluntarism of the state and its unwillingness to 'provide accessibility' in the built environment, yet it also relates to the difficulties that people with disabilities have had to confront in contesting the centrality of the neo-liberalist strategies of the state.

In particular, significant barriers have prevented the emergence of strong, politicized, disability movements and, in general, even in the USA where disabled people have been highly active, it seems that people with disabilities do not act collectively 'for themselves'. As chapter 1 has indicated, people with disabilities tend to be the least powerful members of society, predominantly working class, with lower than average educational attainments and incomes. Such circumstances, as Shakespeare (1994) has noted, tend to preclude effective political organization and action. Moreover, disabled people also spend more time in the company of 'able-bodied' individuals and, as Scotch (1988, 1989) has argued, this experience of disability is often 'individualizing'. For instance, many disabled people are confined to their homes where often their only contact is with an 'able-bodied' carer and immediate (able-bodied) members of their family. Such day-to-day living contexts are potentially inimical towards conceiving of disability as social and structural because the situation, where the immediate reference groups do not share the physical and/or mental impairment, seemingly reinforces the normalcy of 'ablebodiment' while perpetuating the notion of disability as somehow abnormal and individual, a product purely of a physiological condition.

In addition, people with disabilities do not cohere in a way that is suggested by the term 'disability'. As previous chapters have noted, the term is chaotic because it seeks to unify divisible, and different, experiences, creating an homogeneity where it does not necessarily exist. Indeed, there is often more that divides than unites people with disabilities, from their different physiological states which produce different experiences of the built, and other, environments, to their contrasting, often institutionalized, domestic and/or living spaces.[4] Also, as Scotch (1988) argues, even where the self identification of disability as a form of social oppression occurs, this does not necessarily mean that group consciousness and/or political action will result. Thus, as Scotch comments,

> disabled people not only lack the common demographic conditions to foster group awareness and activism but the social status of being disabled can create serious disincentives for many to identify themselves as disabled and act collectively on that basis . . . to be perceived as disabled is typically to be seen as weak and helpless (p. 160).

Moreover, while disabled people's organizations have increasingly moved towards social models of disability which emphasize shared experiences or the 'commonality' of disablement, for Priestly (1995), the 'commonality' approach 'raises difficulties for some groups who consider that they have significant separate interests to which the mainstream of the movement is not as yet adequately responding' (p. 157). For instance, as Priestly's (1995) research on the political mobilization of blind Asian people indicates, issues of 'difference' were seen as central in their activism and, as Priestly concludes, 'respondents identified more strongly with their experience of specific impairments and specific cultural identity than with the common experience of disability' (p. 159). In this sense, there is often more that fragments than binds different types of people with disabilities and it is clear that the lived experiences of disabled people do not necessarily create the basis for a unified political organization (also see Oliver, 1990).

Such fragmentation, in part, seems related to the continuing political marginalization of disabled people despite more radical campaigning activity over the last few years. In the context of access, for instance, most of the campaigning has been institutionalized by the state. In 1983, for example, the Access Committee for England (ACE) was set up as an independent policy advisory committee on access, yet it was, and still is, trapped in a voluntarism which permits government to more or less side-line any of its recommendations if it wishes to do so (Barnes, 1991). As Barnes recounts, the Minister for the Disabled (1990) identified the proper functioning of the ACE as 'something quite different from a pressure group, with its members not representing any particular interest or group but addressing issues on an informal basis' (p. 2). Local authorities have also been exhorted to set up access committees or groups to represent disabled people's viewpoints within the overall umbrella of the ACE, yet as Barnes recounts, the power of such organizations seems to be linked to the patronage of the local authorities or of their willingness (or not) to concede some power and devolved control to people with disabilities.

Such formalized systems seem problematical in a number of ways, and one of the basic failures of the ACE is the absence of any clear social and/or political agenda for people with disabilities (because of its direct links with government). Indeed, the ACE's approach to access is a form of technical and/or design reductionism, where the organization tends to concentrate on the symptoms rather than the causes of disablism in the built environment. This is revealed in a range of documents including a recent publication, *Working Together for Access* (1995). The main focus of the document is visual impairment and access and the advice about, for instance, access groups 'influencing politicians' concludes that 'access groups had a role here by contacting their MPs on access issues' (p. 21). There is little sense of a political strategy or approach while the document is suffused with details about tactile walkways and other design configurations in seeking to overturn inaccessible places. While such matters are not irrelevant, such perspectives, in themselves, serve to de-politicize access issues by locating the source of the problems facing people with disabilities in the 'fixtures and fittings' of the built environment rather than in the socio-political structures of society.

Likewise, while there seems to be great variability between access groups, in terms of size, resourcing, and campaigning styles research shows their marginality from the centre of political power (ACE, 1994c). In a wide-ranging survey of 232 access groups in England, the ACE (1994c) concluded that access groups felt under-used by local authorities while two out of every five felt that access issues could be given a higher profile by local government departments. Moreover, frustration was expressed about the lack of consultation, and the resultant 'frustration, duplication of effort, and wasted money' (p. 4). As an access group in North Yorkshire commented, the local council 'tend to go ahead with things . . . and wait for criticisms', while an access group in Cheshire complained that 'planning officers frequently fail to ensure that existing legislation is enforced'. Moreover, the ACE report paints a picture of access groups which suggests that they remain locked into the voluntaristic structures of the state, while constrained by the absence of effective powers by which to influence local authority access strategies (Barnes, 1991).

For Barnes, such groups are limited because they tend to reflect the wider problems facing disabled people in developing an 'active politics': partial representation, lack of resources, the absence of political skills, while remaining wedded to the state for their (effective) power. Indeed, as the next section of the chapter will illustrate, the nature of access politics, where access groups are involved, is reactive or where access groups are dependent upon their patrons (the local authority) for information, guidance, and resourcing. Such structures seem to reflect what Wolfe (1977) calls the 'franchise state' or where access groups are just one of many competing pressure groups seeking to influence the distributive policies of the state. For Wolfe, a feature of the modern 'franchized state' is its orientation towards the institutionalization of competing welfare demands, or where interest-oriented government agencies seek to work and negotiate with the representatives of specific pressure groups. Yet, for Young (1990), Wolfe (1977), and others, the 'terms of engagement' (set by the state) seek to de-politicize and defuse the demands of pressure groups, while refusing to recognize any interest which is not part of the formalized systems of the franchized networks.

As Young (1990) notes, de-politicization occurs because much of the determinate decision making occurs in semi-secret, while pressure groups are given limited access to some of the key committees. Indeed, access groups, responding to the ACE report (1994c), were suggesting limits to their democratic involvement in local government and, as one access group noted, 'a major success would be to get the local authority interested in access by meeting us and responding to our letters' (ACE, 1994c, p. 5). Moreover, the state, according to Wolfe (1977), seeks to (re) define the legitimate concerns of pressure groups by directing their 'interests' to distributive policy issues about 'the allocation of resources and the provision of social services' (Young, 1990, p. 71). As an access group in my postal survey noted, 'we talk to them (the local authority) quite a bit but they say they can't do this or that because they haven't got the resources . . . but, they might be concerned about this, but we're also interested in getting some more control . . . this is our problem'. Politics, then, as Smith and Judd (1984) argue, is conceived by the state as 'the best way to allocate the surplus for individual and collective consumption rather than the more central question of the best way to control the process to realize social needs and the full potentialities of human beings' (p. 184).

THE POLITICS OF ACCESS AT THE LOCAL LEVEL: THE ROLE OF ACCESS GROUPS

Since the Snowdon report (1979), local authorities have been encouraged to help the formation of access groups and to take into account the views of local disability groups in planning for access, yet there is little documentation of how local authorities interact with such groups or of how disabled people's actions more broadly influence the content of access policies. Yet access groups have proliferated since the early 1980s with the ACE (1994c) estimating that 401 exist in England. Most are small scale in terms of staffing, resources, and political leverage, yet apart from the ACE (1994c) report, there is little documentation on them. They represent one mechanism to enable disabled people to cultivate political power and influence yet as Barnes (1991) suggests, the existence of an access group does not necessarily mean that a local authority will respond to their requests. In noting this, he provides the apt illustration of the Leeds Access Committee celebrating its twenty-first year of existence in 1989, in a city where the main public library and museum were inaccessible to wheelchair users and where disabled car drivers were excluded from large areas of the pedestrianized shopping centre (Anderson, 1990; Barnes, 1991). This highlights one of the key issues of concern, that is, whether or not access groups represent a means of empowering people with disabilities, of facilitating forms of political autonomy, or whether or not they solely reflect, as the Royal Town Planning Institute (RTPI, 1993, p. 28) suggests, the best way for local authorities to facilitate their consultative duties.

In gaining a wider perspective on such issues, part of my postal survey asked local authorities about their relationships with any access groups. Of the 120 responses, 102 (85 per cent) authorities said there was an access group in their locale, yet most seemed to concur with the view of one respondent who noted that 'we rarely see the access group . . . they're small scale and don't do much' . Indeed, the last point is, in part, confirmed when one considers the forms of involvement by access groups in some of the crucial elements of decision making in the local authority structures. As Table 7.1 reveals, access groups appear to be confined to tightly defined, prescribed or received activities, that is, involvement in what I would term 'peripheral' as distinct from 'core' levels of influence. Thus, 64 (63 per cent) of the access groups regularly inspect planning applications, while 81 (79 per cent) have regular meetings with the designated access officer. Yet, the inspection of applications, in itself, confers little or no power on the access group because, as a respondent to the postal survey said, 'we're only obliged to listen to the group's comments, not take them on if we don't want to'. Likewise, meetings with planning officers, while potentially useful, maintain a distance between people with disabilities and the executive committees and/or other structures within which some of the more crucial decisions on access are reached.

Table 7.1 What involvement does the access group have in access issues

Does the access group have any involvement in the items listed below	Yes	No
	Numbers (%)	Numbers (%)
A regular inspection of planning applications	64 (62.7)	26 (25.4)
Regular meetings with the access officer	81 (79.4)	16 (15.6)
Representation on the planning committee	8 (7.8)	73 (71.5)
Access to councillors	75 (73.5)	17 (16.6)

Note: Responses are out of 102 authorities who responded by saying that they had an access group operating in their localities. Where responses do not add up to 102 the missing numbers signify a non response. Also, all percentages are calculated from the original figure of 102.

Source: Postal Survey, 1994-95

In contrast, only 8 of the 102 access groups had some form of representation on the planning committee while, as a planning officer noted, 'we tend to keep the access group at arms length if only to allow us to get on'. Indeed, representative structures, based upon consultation, are the typical form of interaction between local authorities and access groups and, according to one authority, 'it is the council's policy to promote, wherever possible, access for the disabled. Meetings are held with the access group and disabled organizations where their comments and recommendations are received and considered on policy matters'. Other respondents reinforced the notion that somehow the access group was, in the words of one respondent, 'to be seen but not heard' while another noted that 'our group is always around here . . . they take up too much of my time'. For this officer, access was a peripheral issue: 'I spend less than 5 per cent of my time on it and only when I'm bothered by the access group'. However, other respondents were less dismissive and, as one respondent argued, 'our access group is vital because they put pressure on everyone . . . without them I don't think much would happen but they've made us think . . . they're always lobbying the councillors and give them quite a hard time . . . however, they are limited in what they can do and they've got to fit in with us'.

This, then, seems to suggest that the activities of access groups are highly channelled and controlled by planning and other professionals, an observation which is supported from other sources (ACE, 1994c; Day, 1985). For instance, Day (1985) recounts the experiences of the access committee in Exeter where specific 'closures' in the local authority were linked to the group's feelings of disaffection. Such feelings were related to the apparent absence of progress on access and the ineffectual nature of the regulatory controls which the local planners were using. Thus, between 1981 and 1985, the access committee checked over 1000 planning applications, yet noted that in only about 8 per cent of the cases did they actually feel as though they had achieved the desired outcome. As the access committee noted, in virtually all applications there had been a failure to comply with the city's code of practice governing access. This situation was explained, in part, as the developers' lack of awareness of the committee's existence and because the planning professionals had, from time to time, failed to pass their comments on to the relevant developers. As Day argues, the problem was compounded by a system which tended to leave it to the developers to contact the planners about access issues:

So few do that it can be rather disappointing. In some cases action is taken without contacting us further. A notable example has been the local hotel which replaced the step in the corridor by a ramp, including a WC facility, and a couple of adapted ground floor bedrooms with a manual hoist available if required. This has been much appreciated by disabled tourists. What a tragedy that they never came back to us to discuss the details of their proposals and so one or two unfortunate little details crept into the modifications which could so easily have been avoided. Not a major tragedy, just an unnecessary one (p. 19).

In amplifying such issues, the rest of the chapter draws on empirical material largely generated by case studies of two contrasting London local authorities, 'Fenway' and 'Renmead' (the names are pseudonyms). The authorities were chosen to reflect diversity and contrasts in their policy approaches to disability and access. In Fenway, for instance, disability has never been regarded as an issue outside of the charitable ethos of 'doing good works' and the underlying conservative political base of the locale has provided little or no support for disability issues. There is no access officer and few resources are given to disabled groups by the local authority. The area is characterized by the local authority's paternalistic, tokenist approach to disability and access, while there is little evidence of any significant political activity by people with disabilities who live locally. In contrast, the local authority in Renmead, underpinned by a tradition of left wing local politics, is highly supportive of people with disabilities and funds a full time access officer while devoting a budget to access campaigning by the local access group. Access issues are interconnected, explicitly, with equal opportunities, and a conception of disability as a state of social oppression underpins the thinking of the key local authority departments. Disabled people are well organized, active, vocal, and have a significant representation on many local authority committees.

Over a period of eight months in 1994, repeat interviews were conducted with a range of key players and individuals in each local authority, including access officers, local planners, councillors, committee and ordinary members of access groups, architects, building control officers, and members of local disability groups (other than the access groups). I also attended the meetings of access groups, and was permitted to examine minutes of these meetings. The discussion is divided into two themes. First, I consider the interrelationships between the variations in local political attitudes and support for access and the ways in which these influence the nature and operations of the contrasting access groups. As the discussion will demonstrate, the operations of access groups are, in part, dependent upon the (received) levels and types of political support which emanate from the local authority. Second, I then look at some of the organizational barriers confronting access groups in cultivating influence and political power.

The importance of politicizing access issues

A significant research lacuna relates to the absence of understanding of what precipitates the formation of access groups, while there is little or no information on how and why they develop specific types of (political) strategies towards access issues. However, in both Fenway and Renmead the relative strengths and degrees of influence of the respective access groups seem to be related to levels of local political support, through the formal political systems but also from the community as a whole. In Fenway, for instance, the local access group described the dominant attitude, of both the local authority and community, as one underpinned by a medical model of disability with the idea that people with disabilities should be, as an access group respondent said, 'grateful for what they get'. As this person noted, 'there's no such thing as a radical politics here, this is a comfortable middle class borough and there seems nothing to get angry or upset about, it leads to a form of quiescence amongst us all, a sort of do-nothing mentality'. Indeed, for this respondent, his rejection of a medical, or individualizing, conception of disability has been related to the general attitudes he faces time and again when he goes out and about on the streets around where he lives. As he commented:

> my gradual involvement in politics came through my anger at a loss of independence. I was angry that my independence was always being challenged . . . it happens all the time and it's worse here than anywhere I've been . . . I mean, every time I try to go to the filmhouse . . . well, they say they're accessible, but the last time it was awful . . . had to phone to say I was coming, I hated this . . . and they had to remove some of the chairs for me to get in. We watched the film and at the end of it I tried to get into the loo. Well, I had to ask for a key and I was just hanging around desperate to go and when I got in there it was being used as a cleaning cupboard and it had the stored seats in there which had been taken out of the auditorium to let me in. I couldn't believe it and they didn't think that I might have wanted to use the toilet.

The respondent interlinked this episode with the wider political ethos of the local council in that 'I'm not aware of any serious party attempt to support disabled people . . . the council hasn't grasped the nettle . . . there are ad hoc groups emerging but it's all spasmodic and a patchwork from the grassroots . . . there's nothing coming from the top'. He also argued that there was a sense in which local politicians were quiescent on most issues, partly because of the relative affluence of the area. As the respondent noted, 'what we haven't got in this borough is a real disadvantaged group . . . we're not disadvantaged enough to raise our opinions'. Moreover, in describing the political attitudes of both councillors and officers in the area, he argued that;

> the council have never been aware of the serious nature of the physical barriers confronting us and when they turned down our request for an access officer to be appointed that was a real blow . . . it is difficult for us to really know who to use in the authority to help us . . . they gave us a booby prize by forming the Access Liaison Committee but it has no funding and is real tokenism. None of us like it and sit on it grudgingly . . . yes, there is a strong element of tokenism in Fenway it makes me angry that there's not the political will to further the aspirations of disabled people . . . council members use organizations like us as a token and don't show by their actions a genuine commitment to further our needs . . . there's no real agenda to address our concerns.

The absence of a radical and radicalizing agenda, one which transcends medical conceptions of disability, has been, in part, fostered by and reflected in the wider socio-political organization of disability politics in the locality hinted at by the respondent. In particular, organizations like the now disbanded Fenway Association for Disabled People (FADP) was, until 1992, the front institution for representing disabled people's concerns in the borough. Yet, as the access respondent noted, as a registered charity the organization reflected its status by cultivating dependency and in no way did it seek to empower its members. As he noted, 'it was run by do-gooders . . . basically people who had time to give to charities but not in a position to analyse what their views or position really was . . . as a charity, like all charities, it was demeaning . . . it spent all of its time raising funds, and it really reflected middle class angst . . . it seemed it was more for them than for us'.

The dissolution of this organization preceded the formation of Action on Disability in Fenway (ADF), which has become the main campaigning organization for people with disabilities in the area. While ADF was originally set up by the local authority to comply with the statutory provisions laid down in the Community Care Act (1992), that is, the requirement of all social service departments to consult with users in the drawing up of community care plans, the organization has moved well beyond its original remit and, as its coordinator argued, 'the local authority didn't think we would grow so fast and proliferate into a whole range of initiatives, they knew we couldn't achieve anything with the funding given and they're surprised that we're looking at other sources of funding'. ADF has a management committee comprising ten non elected members and, as the chair said, 'there's no point electing them because they're all we've got anyway . . . it's just symptomatic of the lack of interest in disability issues. We've got no real programme and I guess we've got to build up some sort of strategic framework, that's the next step.' While the social services department had reiterated to ADF their need to stay within the confines of the remit given to them, of acting as a consultative committee, the organization has developed an access group and, as its chairperson said, 'effectively ADF and the access group are one and the same thing . . . they're both run by the same people which shows the lack of support we get from disabled people here'.

Part of the problem for ADF and, by association, the access group, is their inclusion in and, indeed, dependence on what Shakespeare (1993) refers to as structures of paternalistic domination dressed up as forms of altruistic benevolence. Such dependence is mirrored in ADF's relationships with its sponsor, the Social Services department and, as the chairperson of ADF commented, 'they mean to help us but it's all like a big hand out and they make us feel as though we should be grateful'. Such paternalism also extends to the funding situation and of the inability of ADF to control the small amount of resources at their disposal. Indeed, while ADF received £10,000 from the local authority in 1994–95, all of the expenditure had to be accounted for through the Social Services department and, as the chair of ADF said, 'they really dictated how we should spend it and we weren't happy about this . . . it just seems to trickle through to us'. The chairperson, in amplifying, remarked, 'we've gained the support of the Community and Voluntary Services but they've been a bit paternal and patronizing . . . they've got us grant funding but they won't release the money to us to control directly . . . can you believe it'. Moreover, any expenditure has to be claimed retrospectively and it has forced ADF to consider its own fund-raising strategies to gain what the chairperson has called 'some real independence'.

The formation of the Access Liaison Group, as one attempt to widen the debate on access in the area, has been characterized by the members of the access group as a 'talking shop which most authority officers seem disinterested in . . . it doesn't have any power to do anything at all'. Indeed, it was formed out of the failure to secure a funded access officer's post and was based on a recommendation from the council that it should bring together local authority officers and disabled people to, in the words of one of the councillors, 'spearhead the drive towards an accessible environment in Fenway'. The reality, however, has been different. While the organization meets about once a month, it largely comprises local authority officers and only a single, wheelchair-bound, disabled person. It has no secretarial support and no funding and the involvement of officers is on a voluntary basis. As the disabled member of the organization commented:

> I feel like the token disabled person here and I don't know too much about access . . . I mean, who am I really speaking for, I can't claim to represent anyone other than myself really . . . the problem is that we talk about access but there's not much understanding about it . . . and what good's an organization like this when some of the crucial people never turn up . . . the architects department never sends anyone down here and how often do we see the senior building controls officer. They're the important ones but they don't seem to take it seriously . . . oh, I guess they're busy people, but . . .

In contrast, Renmead has a long standing commitment towards equal opportunities and, as the access officer noted, 'the ideological framework of equal opportunities is crucial to access being fully pursued . . . there is no point in having a commitment to equal opportunities if access is not addressed'. This commitment was first initiated in 1986 through the setting up of the Equal Opportunities Unit, followed by its subdivision into three sub committees - race, gender and disability – in 1987. According to the access officer, 'disability has always been discussed and forms a part of the agendas of all the sub-committees . . . we recognize disability as something which cuts right across race and gender'. In particular, the council has responded by generating a dense institutional network to support access issues. In 1987, for instance, the Renmead Association of Disabled People (RADP) was founded as a campaigning group primarily with the remit of advising the council on the provision of health and social services, access, welfare benefits and education and employment. In turn, an access group was appointed as one of its sub-committees and, as the access officer remarked, 'the committee members are all committed to the empowerment of disabled people . . . we operate with a model of social oppression here and the council seem to go with it'.

This politicization is reflected in the organizational depth characterizing the campaigning work of RADP and, as its coordinator argued, 'we can't just campaign for access without relating it, holistically, to a lot of other issues. We need to attack the sources of oppression facing people with disabilities much more broadly than just digging up the built environment . . . that won't change a lot really'. To this end, RADP have set up a disability harassment group with the remit to campaign on two fronts. The first specifically seeks to increase the levels of housing provision from the local housing authority; the second element is to attack prejudicial values and pejorative stereotypes of the disabled in the borough by setting up public meetings and discussion to try and increase the levels of awareness of what it is like to be a person with a disability living in a world that generally devalues the disabled person. As the access officer said, 'access isn't a separate issue here so we've got to attack the values which generate inaccessible places . . . so we go into the schools to talk to the kids and we're trying to educate the councillors and officers by persuading them to read the awareness training pack that we've just produced . . . at least it's a start'.

The real contrast with Fenway, however, relates to the scale of active volunteer involvement, the level of financial and other support by the local authority, and the higher level of borough-wide politicization of disability issues. As the RADP's co-ordinator said, 'we can rely on a core of at least 20 to 25 core volunteers to do whatever task we want . . . our problem is keeping them satisfied with things to do . . . we've also been given a budget of £34,000 for 1994–95 and we'll use quite a bit of that on access, although some of it goes to pay my salary'. Indeed, the RADP has 450 individual members, all with disabilities, while there are another 150 affiliated, non disabled, people linked to the organization. As RADP's co-ordinator argued, 'we're always at the big campaigns and we're well networked with other groups nation-wide . . . disabled people here are angry and they get involved'. The organization has an active executive committee of twelve people and, as the co-ordinator noted, 'we don't just deal with access, we deal with everything . . . the committee members also represent us elsewhere in the authority to promote awareness of disability . . . so, we've got people on the local Race Equality Council and the Afro-Caribbean Association'.

In cultivating influence in the area, the contrasts with Fenway are revealing in that, for example, representatives from RADP sit on the planning committee and have an input into the final decisions on submitted planning applications. As the co-ordinator said, 'this is vital to us . . . he's intervened on a number of applications at committee stage . . . there was one recently, a shop conversion and no one had picked up the access implications before it had reached committee . . . he spotted it there and got a condition imposed to have a step removed'. The level of direct influence of the group is, however, difficult to gauge, yet some local councillors in the area are committed to a civil rights agenda and, as the co-ordinator of RADP commented, 'they see it as a political issue, they see we need resources, they always come to me for advice and they turn up to help us lobby . . . they've even been involved in direct action with us . . . most recently when a few of the group chained themselves to the railings at Westminster protesting against the Disability Discrimination Bill'.

The barriers to representation

As some research suggests, access groups tend to be localized and dependent upon key energetic people to develop programmes and policies while their absence of funding tends to reduce their abilities to act (ACE, 1994c). Not surprisingly, the range of activities which access groups are involved in is restricted to local issues, while, as the case of Fenway shows, they tend to be reactive and dependent upon social service departments for fiscal and other forms of support. There is also evidence that many access groups survive by virtue of voluntary labour while having not more than consultative status within local authority structures (ACE, 1994c; Barnes, 1991). Such evidence tends to reflect the situation in Fenway and as the chairperson of the access group recounted, they had been going through a bad time, characterized by 'infrequent meetings, nobody taking responsibility to write letters and sort out key correspondences', and no real political astuteness with no one in the organization knowing who 'to put the finger on . . . we haven't got any representatives in the council chamber and we're just not networked . . . we've got someone on the Social Services committee but they don't have a vote'. One of the difficulties that has confronted the group is knowing who to liaise with in the local authority about access and, as a committee member noted, 'every time we go to the planning department a different person seems to see us about access'.

Such situations do not help access groups in trying to penetrate the inner operations of the planning system and in both localities respondents pointed out their difficulties in making sense of planning applications and in providing comments and advice sensitized to the different types of people with disabilities. In ADF, for instance, the person given responsibility for commenting on applications was unequivocal about his position. As he said, 'I'm not coming from a position of expertise . . . I'm not trained around the parameters about the provision of disability facilities and I'm not sure what to look for and not at all certain what I'm supposed to say or recommend . . . it places quite a bit of responsibility on me really . . . I mean, I know what it's like to be in a wheelchair but that's about it'. The respondent was also concerned about the lack of formal procedures and the ad hoc way in which everything was done. As he remarked, 'I never know what I'm going to receive or even when I'm going to receive it and I never really know if they take my views on board'. Likewise, a respondent from RADP was sceptical about the access officer having sole responsibility for looking at, and providing advice on, planning applications noting how his consultations with them did not always occur. Moreover, in both areas, there was some concern that processes were not sufficiently formalized, or, as the chairperson of ADF said:

> What would be better if there was a formal consultation process . . . someone needs to provide training on access and disabled people and have core people responsible for making recommendations about access in submitted planning applications . . . this is the only way to get serious comment . . . what I think as a wheelchair user is not the same for a blind person.

In both localities, respondents from the access groups also commented on the strictures imposed on them by the statutory turnover time of planning applications and the requirement that decisions be made within an eight-week period. As a respondent from Renmead commented, 'by the time we've got around to actually looking at the applications the whole process has nearly elapsed'. Planners in both of the study areas concurred with this and even suggested that their own imposed deadlines on the access groups were potentially prohibiting them reaching an opinion about the submitted applications. As one officer in Fenway remarked, 'well, we give them a few days, but the committee cycle means we can't give them much more time'. Likewise, in Renmead the time scales, as the access officer said, are a problem, 'a few years ago we had about a month to get our replies back, now it's just a few days . . . it isn't easy'. For the access groups, consultation was, inevitably, rushed, and an access group member from Fenway said, 'half the time I can't look at all the applications because of the deadlines I'm supposed to work to . . . they seem to want them back straightaway and it doesn't give me much time to talk with some of the other members'. On a few occasions the respondent had not been sent the weekly batch of applications, while many had only turned up at the last minute, giving him little time to respond within the deadlines set by the local authority. As the respondent said, 'if I don't respond, they don't get my view and I guess they just think I haven't got anything to say about the batch they've sent to me . . . well, it's not really like that, is it'.

There is also the issue of who sets the agenda on access issues and as a member of Fenway's access group commented, 'we don't really have any input into setting the agenda on the Liaison Group and we never seem to be asked to say what we think are the real issues . . . what's the point in going along'. Other members of the group were also cynical about their inclusion in some of the local authority committees and, as the chair of the access group said, 'every time we turn up the agenda's just there waiting for us . . . it's all handed down and we're expected to get on with it . . . the trouble is that we do and we're not very good at making a fuss about it'. Such estrangement of the group has also been reflected in some of the day-to-day interactions between them and the local planning authority. For instance, the group receives planning applications on a weekly basis to look at yet, as the chairperson commented, 'we never get to say which ones we'd like to see . . . they just turn up'. Moreover, as one of the planning officers in Fenway responded, 'it's our job to identify the relevant from the non relevant applications and then we'll present a selection to the access group . . . we can't really let them get in here amongst the files'. However, in Renmead, the access officer meets with planning officers on a weekly basis with a disabled representative from RADP to choose which applications to scrutinize. As the access officer said, 'we have complete discretion about what we look at and we can take any application away . . . we can also come and go as we please, but we do have a good relationship with the officers'.

One of the key problems facing ADF relates to the organization of its meetings and the almost insurmountable difficulties of its members in getting to a common rendezvous point. As the chair has argued, 'all of us are mobility impaired one way or another and a couple of the group find it more or less impossible to leave their homes . . . that's our first problem, just getting here to meet'. While the members of the access group in Fenway are more or less dependent on their own, individual, means of getting to meetings, the contrast in Renmead could not be greater. As RADP's co-ordinator noted, 'we can get people here by one of two means: we either use dial-a-ride services by booking the service in advance and RADP pays all of the costs; or, we can borrow mini-buses from some of the other voluntary associations in the borough and they'll be driven by volunteers and all the vehicles have tail-gates . . . so, we never have any problems getting a good turnout of members here'. In Renmead, the access group is fully supported by a full time secretary and office space, while their identity is reinforced by the provision of items like letter headed note paper. However, in Fenway, although the ADF was set up by Social Services to satisfy the provisions of the Community Care legislation, it receives little or no support from the local authority. As the chair of the group commented, 'how can we be effective if they don't give us support . . . we have no secretarial support, nothing for stationery, and we can't claim travel expenses . . . it stops us doing a lot of things'.

Another problem experienced by access groups was not knowing who is disabled or where they live. As Renmead's access officer commented, 'our initial strategy was to find out where disabled people are in the community . . . we don't know . . . so we go and talk at day centres and we use the education department to publicize that we're around and can help'. Indeed, part of the difficulty for the access groups, in reaching beyond their core membership is, as the co-ordinator for RADP commented, 'the lack of shared experiences between the different types of disabled people . . . we all experience the world differently and it's difficult to find some common ground'. In particular, the fragmented, often disparate, nature of the membership of the access groups in both areas was commented on by interviewees and, as the chair of the Fenway access group remarked, 'we do not speak with a concerted voice. Our activities are too individual and we require more co-ordination . . . and we don't appeal to enough people'. Indeed, as one of the ADF representatives said, 'a problem for us is just contacting disabled people out there, letting them know we're here, but then trying to get them to join . . . we don't seem to appeal much especially to the youth'. In particular, the co-ordinator of ADF related the apathy with disability issues in the area to the absence of support structures and, as he commented, compared to Bethlay (where I used to work) there is no culture or climate of voluntary organizations here . . . anything that's created tends to fall apart in a short space of time . . . ADF might suffer the same if we don't get an increase in membership, and also members who just aren't in wheelchairs'.

In both localities there was a concern with the domination of physical impairment as the defining benchmark for access and, according to the co-ordinator of RADP, 'it's one of the major issues to be faced because it is exclusive and wrong and will never really address the concerns of the majority of people with disabilities'. An illustration of such bias is Renmead council's Access Awards Scheme, an annual competition to recognize and reward the most accessible buildings in the borough. Yet, as RADP's co-ordinator commented, 'a couple of years ago our building won one of the awards but the first thing I noticed when I arrived here two months ago was how it was only really accessible to wheelchairs . . . it's got a nice ramp in but that's about it'. In responding to this, the co-ordinator has been trying to transform attitudes in the RADP towards incorporating those previously underrepresented and/or excluded from the main campaigns. As the co-ordinator noted, 'people with learning difficulties and sensory impairments seem to lack any voice here . . . when access as an issue is talked about it never seems to be about anybody other than someone with a physical impairment . . . we've got to do something about this'. Of key concern for the co-ordinator are people with severe learning difficulties and/or individuals with mental impairments which do not permit them to communicate in conventional ways. As the co-ordinator said, 'just because they can't talk or say anything verbally doesn't mean that they can't communicate yet a whole load of people out there are being discriminated against because they can't shout or say things in the classic or conventional way'.

Yet, for the RADP there is a potential crisis of identity in that those who are activists within the organization are overwhelmingly the elderly and/or those with a physical disability and, in the co-ordinator's terms, 'lacking empathy towards the wider community of disabled people in the borough'. Indeed, the committee of the RADP comprises sixteen people, the majority of whom have a physical impairment and, as the co-ordinator noted, 'it just reinforces the stereotype that most people hold of disability'. In trying to change the committee structure, however, the co-ordinator is aware of the longevity of its existing incumbents and of the potential resistance to sudden and/or dramatic changes. For her, the political poser is how to have a stronger and more varied input into the way the association is run while trying to get different types of disabled people onto the executive committee with 'the right kind of support'. As one of the committee members commented, 'the problem we face is that if we claim to speak for disabled people we need to attract a varied membership'. Likewise, in Fenway, the chairperson of the access group felt that there was a problem of legitimation because they only had twelve members in total and most of them in wheelchairs. As he commented, 'when it comes to access issues we still can't say we are the voice of disabled people here'.

CONCLUSIONS

The cases illustrate that the degree and type of institutional support by local authorities for access groups is crucial for conditioning the ways in which the latter are able to operate. Yet, as the ACE (1994c) has reported, there is much variation in the support given to access groups by local authorities and, in the case of Fenway, the main campaigning group, the ADF, was set up primarily to fulfil the requirements of community care legislation. As ADF's co-ordinator argued, 'we know we were set up to satisfy them (the local authority) but we're trying to make the best of it'. Indeed, variations in local government support for access are, not surprisingly, interrelated to the wider socio-political values of the areas within which access groups operate. Thus, in Renmead, a wider civil rights agenda has, historically, emerged out of a poor ethnic and working class population. By the late 1980s, when access became an issue, campaigners were able to merge into a pre-existing political infrastructure which was committed, in part, towards policies of equal opportunities for marginal groups. In contrast, Fenmead's affluent population has underpinned an ethos of voluntary, charitable, works, and where levels of politicization have rarely transcended the stage of, as the local access co-ordinator described it, 'the do-gooders good works mentality'.

The roles of the access groups in both places also need to be understood in relation to the nature of local politics within the local authority structures. Thus, in Fenway, the access group has little or no representation on key committees within the local authority, while its resource base is tightly delimited by the Social Services department. Moreover, Social Services in Fenway tend to conceive of disability in biomedical terms and see little need for (political) campaigning. In contrast, the local authority in Renmead conceives of disability as a form of social oppression and the access group is encouraged to campaign against the national government, in effect, to politicize access as a function of transforming the wider (national) political base. Funds are provided for such purposes. Ultimately, however, the access groups in both areas remain tied to their locales and constrained by the absence of mechanisms to co-ordinate and propagate access politics at the national level. Indeed, as previous chapters have suggested, the extent to which significant change can occur locally is constrained by the nature of the legislation governing accessible environments. This concurs with the ACE's (1994c) survey of access groups where the observation by an access group in Devon was typical, in that 'we need more national co-ordination of access groups . . . a clearer structure . . . reinforcing a national identity'.

In the UK, disabled groups have yet to clearly articulate a politics of difference or to assert their self defined agendas which are not part of integrationist policies. The ACE (1994c) survey, for instance, indicates a low level of non strategic activity by access groups, so too my postal survey, which shows most of the groups involving themselves in activities prescribed, or handed down, by their respective local authorities. Access groups are also fragmented and, although Fenway and Renmead border each other, neither access group had heard of the other (also see ACE, 1994c). The two cases also show, in part, a semblance of Wolfe's (1977) conception of interest group politics as part of a franchised state, although this is more evident in Fenway than Renmead. Moreover, while the politics of access, in both places, have been constructed around a range of assimilationist ideals defined, largely, in and through the state, in Renmead there is a keener sense of the self assertiveness of people with disabilities, of trying to define agendas and strategies not necessarily handed down by the local state. In interview, for instance, a member of the access group, with cerebral palsy, commented how 'it's just a difference, we're not the same as you . . . we've got to use that fact as our strength, celebrate the difference and fight to assert it'. In Fenway, by contrast, there was more a sense of pragmatism, of trying to establish disability *per se* as an issue within the local authority than of seeking to encourage variegated policies about the multiple meanings and forms of disablement.

The 'politics' of access, as illustrated by the case studies, indicates variety and possibilities yet clearly highlights significant organizational, practical and resource constraints facing access groups. Moreover, there are wider questions and issues to be broached beyond the confines of the study material reported here. In particular, is it possible for local groups to achieve any measure of gain by remaining 'localized'? This, then, is a question of the appropriate spatial scale in and through which the politicization of access should (and could) occur. For access groups, the only national co-ordinating arm in England, for instance, is the ACE and they have been castigated as an organization that only consults access groups 'part way through issues, rather than the start and results are rarely published'. Other groups have also exhorted the ACE to 'draw from them' and enable them 'to pursue' their 'own objectives' (ACE, 1994c, p. 5). Indeed, there is a sense that access groups remain weak and isolated, 'fire-fighting' for access in a wider socio-political context that does not understand, or even want to know about, the oppressive conditions of the built environment. For the access co-ordinator in Fenway, 'it will be a long time before I can ever take myself out of my house . . . for eight years, I've only been able to go out because of my carer . . . it's frustrating, it's not right'.

NOTES

1. Figures for 1992 estimate that 22 per cent of employers overall are meeting the 3 per cent jobs quota figure (Summers, 1992). Summers argues that only one government department out of a total of 29 was meeting the 3 per cent quota for the employment of registered disabled people in 1992. The figures indicate that 3.2 per cent of the Department of Employment's workforce was registered disabled while the overall

average for government departments was 1.4 per cent. Summers also notes that only 12 district councils in England and Wales out of 366 and none of 47 county councils or 33 London boroughs were making the quota.

2. The more militant actions and political demands of disabled people indicate that equal opportunities and treatment cannot really occur and would, in any case, be contradictory in terms of what people with disabilities are seeking to achieve: that is, the rights to define their own lives and live them in full celebration of a recognition of their differences. Indeed, the whole issue of equal opportunities in planning, and elsewhere, really has to come to terms with what Birkenbach (1993), and others, have referred to as the 'dilemma of difference' (also see Young, 1990). Goffman (1963) recognized the dilemma in relation to the stigmatizing of disabled people and highlighted the essentially self defeating options available to disabled people and policy makers alike. These options are, on the one hand, seeking integration with the 'mainstream' yet, in doing so, drawing attention to the stigmatizing difference, or, on the other hand, maintaining segregation which, in turn, only serves to legitimize the pejorative labels attributed in the first instance.

3. Disabled people in Britain have been organized into pressure groups since the late nineteenth century, starting with the National League of the Blind and the British Deaf Association. See Pagel (1988) and Driedger (1989) for accounts.

4. Such circumstances, then, are implicated in the disparate attitudes and (political) responses to access issues and, as a range of authors has documented, people with different types of physical and/or mental impairments necessarily relate to and use the built, and other, environments in all sorts of different and often mutually conflicting ways. For instance, for people with specific types of visual impairments, perception of depth and/or distance can be difficult and it can be more or less impossible for individuals to easily negotiate steep stairways. For other types of disabled people, such as a person who is hard-of-hearing, such stairways would pose no difficulty for them. To the extent that such differences are, in part, determinate of disabled people's attitudes towards access, it seems likely that the emergence of a specific (universalized) politics in and around access issues is likely to be constrained.

FURTHER READING

There is a range of descriptive accounts of the disabled people's movement, including Pagel (1988) and Driedger (1989). While useful accounts they tend to be brief and lack any sense of the oppressive status of people with disabilities. Shakespeare's (1993) article is a much more theorized, and thoughtful, account of the emergence of disability politics precisely because he situates the rise of disabled pressure group politics in the context of what he terms 'liberation' politics.

8

Beyond Disabling Environments

As a blind man he could rise in the morning, help get the children off to school, bid his wife good-bye, and proceed along the street and bus lines to his daily work, without dog, cane, or guide, if such his habit or preference, now and then brushing a tree or kicking a kerb, but, notwithstanding, proceeding with firm step and sure air, knowing that he is part of the public for whom the streets are built and maintained in reasonable safety, by the help of his taxes and that he shares with others this part of the world in which he, too, has a right to live.

Testimony of tenBroek, 1966, p. 1

INTRODUCTION

tenBroek's testimony, of a world without prejudices against people with disabilities and of environments free from physical and social barriers, remains a distant and probably unachievable ideal. For most disabled people, their societal marginalization is an established facet of their everyday lives, from the continuing dominance of a demeaning charitable ethos to the pejorative images which the media transmit about them. The testimony also stands at odds with the wider socio-political structures of western welfare states and, as Barnes (1991) has argued, their legislative bases relating to access, and disability more generally, reinforce 'disabling environments' so perpetuating patterns of apartheid between the 'able-bodied' and the 'disabled'. In the UK, for instance, governments have refused to legislate to make domestic dwellings accessible to all types of disabled people, while the provision of accessible transport is a discretionary matter for local authorities. Raised kerbs into shops are placed everywhere, induction loops are rarely found in public spaces, while tactile walkways for the visually impaired rarely feature as an integrative and integral component of the built environment. As an access officer for a London local authority commented in interview, 'everytime I go out it's like an assault course, I can hardly get down the street because the paving stones are shattered, and I feel that everyone's looking at me . . . I can only go to certain places on my own . . . it's a form of enforced dependence'.

While tenBroek espouses the right to live in a world free of physical barriers, the period since the early 1980s, particularly in the UK, has witnessed a curtailment of disabled people's rights with a general movement from statutory based services to voluntaristic codes and practices (Barnes, 1991; Imrie, 1996c). In part, this reflects a wider truncation of welfare services in a context in which welfare policy has been shifting from collectivist to individualist, or market-based, forms of provision. Thus, as the present government (Disability Unit, 1994) has acknowledged, in consultation with a range of groups about the contents of the Disability Discrimination Act, service providers and businesses will not be compelled to facilitate accessibility. Rather, as the government has argued, 'disabled people may encounter discrimination . . . the government has worked hard to change these attitudes through a continuing policy of education and persuasion . . . more direct measures have been taken where they are practicable and do not place an unreasonable burden on business' (in Laurance, 1995, p. 3). This, then, prioritizes the (self defined) needs of the development industry, while indicating that the provision of accessible places will be largely dependent on disabled people expressing some form of market demand for them.

The 'marketization' of access has also been reflected in the directives which government has issued to local authorities, and a recurrent aspect of local planning, since the early 1980s, has been the emphasis on the need to enhance the speed and efficiency of the system and for planners to be responsive to their clients. Yet, perversely, the interpretation of 'client' is partial and biased towards the development industry and its requirements while little attention has been devoted to the end user. The planning system is also operating with weak and reactive legislation and, as chapters 5 and 6 have indicated, planning policy advice from the Department of the Environment (DoE) throughout the last ten years has encouraged planners to 'facilitate' development while avoiding prescriptive policies and/or planning conditions which are not strictly related to physical land use matters. In this sense, planners have been encouraged to operate within a technical-procedural conception of the planning process, while de-emphasizing the social and ethical content of planning procedures and/or outcomes. In turn, this has encouraged a 'design reductionism' towards access, a form of environmental determinism.

Moreover, the reality for most local planning authorities, as for local authorities as a whole, is one in which they remain remote from their constituents, and where the representative structures of governance do little to encourage the participation of user groups in the design and implementation of policies. Access groups, for instance, rarely participate in the detailed decision-making of development control, while planners tend to operate with conceptions of disability as somehow a uniform 'condition' to be 'treated'. In seeking to transcend such thinking, Healey (1992) makes a case for an environmental planning system which is increasingly 'market sensitive and accountable to the wide range of environmental interests' (p. 428). The issue, of course, is how to translate the changing substantive matter into new practices, procedures and policies. While politicians and others are stressing the integration of economic, environmental and social objectives, one must concur with Healey that one of the main challenges is to transform the planning system from a 'hierarchical, techno-bureaucratic practice' to one with greater levels of local choices and decision making. This also means a greater concern with individual liberties, a genuine democratization of the system, and an attempt to go beyond Citizens' Charters towards building a real sense of public involvement in strategic, and other, decisions. This, then, is a call for participative democratic structures.

It also seems important to offer alternative possibilities for people with disabilities or, as Marcuse (1964) has argued, to 'conceptualize the stuff of which the experienced world consists . . . with a view to its possibilities, in the light of their actual limitation, suppression, and denial' (p. 7). People with disabilities have already started to (re) conceptualize and transform the worlds that they inhabit and the emergence of direct action is, in part, premised on an assertion of their differences, and on critiques of state welfarism and the paternalism of the wider structures of the 'caring' industries. As Barton (1993) has argued, it is time for disabled people to 'overthrow patronizing and disabling expectations and practices and challenge notions of passivity and dependency' (p. 246). Yet, for Young (1990) and others, such possibilities are still to be translated into specific political structures or ways of living which revolve around a (re) centring of (bodily) difference as something to be celebrated rather than denigrated. In particular, part of this must relate to breaking down the estrangement between 'the abled' and 'the disabled' and of transforming the asymmetrical power relations within which people with disabilities are embedded. Indeed, the subjectivities of people with disabilities are still systematically denied to them and it seems difficult to disagree with Barton's (1993) conclusion that nothing short of well organized protest and civil disobedience will begin to challenge the disablist values and practices of society.

I wish to extend some of these themes and, in particular, to consider some of the possibilities for people with disabilities in seeking to transcend the ableist socio-political practices and policies of state and society. In the next section, I consider some of the ambiguities, problems and possibilities of anti-discrimination, or equal opportunity, policies in the planning system. Over the last ten years, planning practice, like most professional services, has become infused with notions of 'best practice', 'serving the consumer', or pursuing policies which promote equality of opportunities for the clients that it serves. Yet, as I shall argue, such debates, while important, tend to ignore questions of how to tackle wider systemic forms of oppression against people with disabilities. In a second section, I broaden the focus by considering conceptions of citizenship, while discussing some of the elements which could contribute towards the political empowerment and emancipation of people with disabilities. In a final section, I relate themes of empowerment to one of the contexts within which disabled people are estranged, that is, the social relations of research production or one of the means by which the subjectivities of people with disabilities are 'revealed' to wider communities.

DIVERSITY, DIFFERENCE, AND THE LIMITATIONS OF
THE ETHOS OF EQUAL OPPORTUNITIES.

A feature of the British planning system is its concern with the pursuit of equal opportunities and with seeking to redress what the Royal Town Planning Institute (RTPI) considers to be discriminatory practices against minority groups. As the RTPI has acknowledged, planning practice is neither neutral nor benign in its effects on communities; planning outcomes have particular distributive consequences which, potentially, exclude some while conferring advantages on others. For Gilroy (1993) equal opportunity policies were first stimulated by the Race Relations Act (1976) which encouraged local authorities to promote equality of opportunity and good race relations. Local authorities were also encouraged to transform potentially discriminatory procedures and practices while opening up services to groups previously excluded from particular forms of provision. In turn, this has provided the planning system with its principal, and defining, focus in seeking to overturn the marginal status of disabled people within the built environment.

Yet it is not clear from Gilroy and/or RTPI statements on equal opportunities what equality really means. For Birkenbach (1993) equality of opportunity is a complex and potentially differentiated concept which is open to many contested interpretations, yet one gets little sense of this from debates (or their absence) within professional planning circles. Thus, as Birkenbach queries, is the profession referring to the means or the ends of equal opportunities? For instance, is it the case that all of us have equal rights to succeed in whatever it is we wish to do, and should the means to do so be provided for us? Or, do we all have the right to an equal chance of success in whatever it is we choose to do? This, then, begs the issue of whether or not equality of opportunity is also about equal outcomes for participants in the planning process and/or whether or not such scenarios are desirable anyway? The underlying definition of equal opportunities in planning is also characterized by seeking to include people in the planning process, yet what is it, in Gilroy's (1993) terms, to be 'excluded from the system' when to be 'included' often involves little more than being locked into forms of representation which do not provide effective means to enable groups like disabled people to influence the actions of professionals?

In this sense, equality of opportunity is rarely equated with breaking down the institutional contours of the planning system to facilitate participative democratic structures, nor does it necessarily entail a demystification of the 'expert' systems which underpin the practices of planning. The RTPI also presents disablism as an injustice which can be largely addressed through the pursuit of clientism or systems of advocacy, yet this fails to bring into focus issues of the domination of people with disabilities by the patterning of institutional organization and decision making, so crucial to the perpetuation of ableist values and practices. For instance, the role of access groups is highly circumscribed by the institutional structures of local authorities, with their ability to act largely dependent on power being handed down to them. Indeed, local planning departments provide access groups with little more than a representational involvement in decisions about accessibility, while largely controlling their resource base. In turn, the particular conception of equal opportunities which the RTPI have in mind is one which leaves intact the hierarchical relations of institutional power of the type which exists between local planning authorities and access groups.

Moreover, the RTPI's propagation of and support for defining equal opportunities as an issue about eradicating discrimination tends to individualize the situation with which people with disabilities are confronted.[1] As Young (1990) argues, discrimination is an agent-oriented, fault-oriented concept in that it 'focuses attention on the perpetrator and a particular action or policy, rather than on victims and their situation' (p. 195; also see Freeman, 1982). There is also the problem that equating a group-based injustice with discrimination depends on the 'victims' demonstrating, on a case-by-case basis, that they have suffered. In turn, this situation tends to minimize the numbers of complainants, often because they are uncertain of the substance of their complaints and of how to use the court systems to their advantage (see Gooding, 1994). Thus, in the context of sex discrimination in the UK, fewer than twelve successful cases, since 1976, have been won in the courts. Moreover, in assigning fault, discrimination tends to conceive of the injustices that people with disabilities have to suffer as, in Young's (1990) terms, 'aberrant, the exception rather than the rule' (p. 196). For instance, since the passing of the Americans with Disabilities Act (1990), many have assumed that the normal situation for disabled people in the USA is the absence of discrimination, yet this is far from so (Imrie, 1996c).

Indeed, part of the difficulty with the RTPI's focus on equal opportunity policies, as one of the ways of eradicating discriminatory practices against people with disabilities, is the underlying premise that discrimination is the principal way in which disabled people are unjustly treated. Yet, as Young (1990) has argued:

> we should deny the assumption, widely held by both proponents and opponents of affirmative action, that discrimination is the only or primary wrong that groups suffer. Oppression, not discrimination, is the primary concept for naming group-related injustice. While discriminatory policies sometimes cause or reinforce oppression, oppression involves many actions, practices, and structures that have little to do with preferring or excluding members of groups in the awarding of benefits (p. 195).

Oppression, then, is multiple in form and, for instance, may be the averted gazes or the cultural stereotypes held of particular groups. It can also relate to forms of institutional domination, like the incarceration of people with disabilities into special institutions or withholding their rights to vote in a general election because a doctor refuses to sign a form to signify that 'they're able' (see Barnes, 1991). Oppression is also embedded in institutional decision making and in the systems which exclude disabled people from influencing decisions which condition the ways in which their lives are led. Indeed, for people with disabilities, while distributive matters are of concern, the concept of oppression is the pivotal category of political discourse and, for Barton (1993) it defines the position of disabled people as one of a situation of both powerlessness and voicelessness and where they need to address 'how dominant groups either ideologically disparage or ruthlessly deny the humanity of the Other' (quoted by Barton, 1993, p. 243, taken from Giroux, 1992, p. 33).

Thus, as Barnes (1991) and others have argued, what people with disabilities want, first and foremost, is social equality, or what Young (1990) defines as the 'full participation and inclusion of everyone in society's major institutions, and the socially supported substantive opportunity for all to develop and exercise their capacities and realize their choices' (p. 173). For disabled people, such capacities and choices revolve around their desires to be treated without disdain, for the pejorative cultural stereotypes to be disbanded, and for their bodily differences to be accepted as something which is not extraordinary. Moreover, many people with disabilities think of distributive issues, such as their inability to gain a job, or an accessible house, as interconnected with systemic forms of socio-cultural oppression which, as Barton (1993) argues, cannot be overturned by (re) distributive equal opportunities policies alone. Thus, in response to the proposed extension of Building Regulations on access to housing, the House Builders Federation (1995) has commented that 'if a disabled person visits a home owner, it is to be expected that they can be assisted over the threshold' (p. 1). This is a reaffirmation of a (cultural) stereotype, of the wheelchair user as a dependent who does not mind, and it serves as a powerful ideological propagator in reinforcing the resistance of the housing building industry to acceding to the pressures to create accessible dwellings.

While policies and programmes premised on equality of opportunity are a component of the pursuit of social equality, they in themselves cannot guarantee that social equality will be achieved for people with disabilities. This is because, as noted above, equal opportunity policies are primarily concerned with the distributive issues of who gets what rather than with challenging the wider socio-political, cultural and ideological structures of society. Moreover, the RTPI, in propagating equality of opportunity in the planning process, has said little or nothing about aligning the disadvantaged positions of people with disabilities to a wider political agenda, with, for instance, conceptions of civil rights and citizenship.

ON POLITICAL EMANCIPATION, ACCESS, AND DISABILITY

As Barton (1993) has argued, the social and economic policies of the last decade in the UK have led to a situation whereby people with disabilities are experiencing a reduction in their degree of choices over goods and services with 'an increase and intensification of scrutiny and control by professionals and others' (p. 242). The reduction in accessible new build housing is one illustration of this wider trend, while the emergence of non elected governmental service providers has introduced an ethos of 'political closure', or the non disclosure of their records of meetings or details or justifications of strategy to the general public. For Barton (1993), Oliver (1990) and others, such trends represent an extension of a market ideology or where the distributive justice of the market is being extolled, a belief in the importance of competition in improving the quality and outcome of services. Such competition is increasingly being encouraged by decentralizing services to a range of private sector organizations, with the expectation that provision of services will reflect the (monetary or market) demand being expressed for them. For Young (1990) the emergent logic is a belief in equality of opportunity to become unequal and, as a range of commentators has observed, the dismantling of systems of universal welfare provision within the state is exacerbating income and material differentials between the rich and the poor, while denying many people in need the access to specific goods and services which they require.

This is directly affecting people with disabilities who have little or no economic power and, increasingly, represent a 'social residual' with few means to exercise market choices. As Barton (1993) notes, the systemic, structural, conditions of disability are a denial of citizenship or, more accurately, the imposition of a conception of citizenship which narrowly revolves around the notion of consumerism. Thus, for successive Conservative governments in the UK, their vision of citizenship is one of 'the active consumer', a purposeful individual taking responsibility for their own lives, providing for themselves and exercising choice (their 'citizenship') through their purchasing power. This, though, is a vision of citizenship which fails to recognize the systemic social inequalities in society, or how the socio-cultural and institutional fabric of society serves to exclude or to mark particular people out as 'the other'. There is, then, a need for people with disabilities to (re) capture conceptions of citizenship. However, citizenship is a complex, contentious, and contested, conception yet, as Marshall (1950) has argued, it must involve the extension of political, social and civil rights to all, albeit within a framework underpinned by individuals accepting that they have responsibilities and obligations.

For Marshall (1950), then, citizenship is defined as comprising the rights necessary for individual freedom, including the liberty of the person, while the right to participate in the exercise of political power is seen as key in empowering people. In addition, the right to economic welfare, security and 'to live the life of a civilized being according to the standards prevailing in the society' is a core element of Marshall's conception (p. 11). Others concur with this more holistic approach in defining citizenship and, for Young (1990), a pre-condition of citizenship is the dismantling of representative forms of governance in favour of participatory citizenship or a group involvement within the institutional fabric of society. As she claims, this is more likely to embody democracy because it 'nurtures such publicity by calling for claimants to justify their demands before others who explicitly stand in different social locations' (p. 190). For access groups, for instance, this would entail their formal participation in the committee structures of local government and precipitate, for example, a direct engagement between them and service providers. Indeed, for Gramsci (1971), such engagements are only transformative and empowering if a 'praxis of the present' is developed, or where active and reciprocal relations between progressive groups and service providers are established, in order to enable the interacting groups to become more conscious of their own actions and positionalities within the world.

Likewise, others note that empowerment can only be achieved by the engagement in 'discussions within and between discursive communities, recognizing, valuing, listening, and seeking for translatative possibilities' (Giddens, 1991, p. 91). Indeed, following Giddens (1991), it seems useful to define emancipatory politics as 'a generic outlook concerned, above all, with liberating individuals and groups from constraints which adversely affect their life chances' (p. 40). As he notes, such a politics has two definable elements: firstly, to overcome illegitimate domination of some individuals or groups by others and, secondly, to create the conditions for a transformative attitude towards the future. As Giddens (1991) argues, the pursuit of an emancipatory politics is about 'self actualization, of (re) gaining control over one's material and psychological conditions of existence, of shedding 'the shackles of the past and overcoming the illegitimate domination of some individuals or groups by others' (p. 211). In developing Giddens' perspective, Oliver (1992) and Zarb (1992) note that the three fundamentals upon which an emancipatory politics should be based are reciprocity, gain and empowerment. For Giddens (1991), for instance, reciprocity implies some give and take between different communities, where mutual negotiations of power take place. This, though, means attempting to transcend the 'normal' motivations of reciprocity pursued by policy makers and/or service professions, that is, of primarily seeking to develop reciprocal relations as a means of controlling user, or client, populations.

As Oliver (1992) notes, reciprocity is difficult to achieve, often because professionals never seem to reveal as much about themselves as they expect to have revealed to them, while relations of trust and interaction between them and their clients, such as people with disabilities, are not easily developed over the short time spans dictated by policy cycles. Moreover, while an emancipatory paradigm seeks to empower people with disabilities, this becomes problematical while the existing institutional (or material) relations of state welfarism persist. In addressing such issues, McAllister (1980) notes that a broader philosophy of professional-user relationships must emerge, based on what he terms citizen participation, a process which simultaneously deconstructs the mysticism of the expert systems, while providing all people in a community with the opportunity to influence. In part, this reflects a wider debate about consumer-based approaches to policy making, yet it goes further in calling for the devolution of power from hierarchies and/or developing power bases which exist in local nodes and networks (also see Brooks and Gagnon, 1991; Jenkins and Gray, 1990). Indeed, as Jenkins and Gray (1991) suggest, most forms of policy say little about service quality and tend to reinforce the distinction between policy makers (the experts) and communities (lay people) by only being attentive to customer relations and systems of managerial responsiveness. In short, the possibilities of policy frameworks incorporating measures of (self defined) community and consumer benefits are limited to managerial or procedural restructuring.

However, as Hammersley (1992) argues, in developing a critique of emancipatory politics, much emancipatory inquiry is based on a simplistic notion of practice. Hammersley argues that those propagating such a politics tend to see the struggle for emancipation from oppression in dualistic, even reductionist, terms, a world neatly divided into oppressors and oppressed. Yet, as Hammersley notes, there is no single type of oppression or oppressor, and he suggests that it is possible to classify different people as oppressors or oppressed from different points of view. Indeed, Hammersley considers the concept of oppression to be problematical in that it assumes that one is able to identify easily what are real and genuine needs. As he notes, 'if oppression is not an all or nothing term – not restricted to a single dimension, it is not clear that it can be overcome in the form of a total, once and for all, release from constraint that the term emancipation implies' (p. 189). Yet, his position represents a 'do nothing' scenario based on relativist claims which, ultimately, only serve to reinforce (and even legitimize) power differentials, and the exercise of that power to the detriment of specific groups in society. In this sense, oppression exists and cannot be ignored. Moreover, emancipation is not necessarily an objective which can be defined solely in terms of a political project conceived of as a 'total, once and for all, release from constraint', as Hammersley implies. Indeed, the incremental gains, and losses experienced by, for example, feminist movements illustrate the possibility of multiple political strategies, while indicating that emancipation, in seeking to break down oppressive social relations, is not easily achievable in the short term. This historiography, then, provides a message to people with disabilities where the failures of policy require a new agenda, of participative democracy and the self definition of problems and policies.

RECONSTRUCTING RESEARCH AGENDAS AND THE SOCIAL RELATIONS OF RESEARCH PRODUCTION

The self definition of problems and policies is one of the core issues of social research and of how knowledge should be generated about the material conditions of disabled people's lives. One of the mechanisms of social oppression is the production of research and/or knowledge which denies the subjectivities of the research subject, and/or where research subjects are objectified as phenomena to be 'gazed at' and to be 'worked on' as though they are an inert and passive substance. Yet, this is how much social research is conceived, and for many people with disabilities the writings and research findings about them often bear little relationship to how they themselves see and experience their lives. There is, therefore, often something of an 'interpretative gap'. Oliver (1992), for instance, powerfully argues that research on disability has had little influence on policy and has made no contribution to improving the lives of disabled people. In particular, he notes how the process of research production has been alienating both for people with disabilities and for researchers themselves. This echoes a wider critique from feminist scholars of research methodologies cast within positivist and interpretative paradigms (Imrie, 1996a; Lather, 1991; Stanley and Wise, 1993). Such critiques locate the sources of oppression, exclusion and, ultimately, misrepresentation of a range of social groups within a specific set of what Oliver (1992) terms the 'social relations of research production'.

Such social relations tend to reflect dominant power groupings in society, and, in the research context, perpetuate an elitist structure which conceives of the researched as somehow subordinate (or inferior) to the researcher. As Oliver (1992) notes, this is fundamentally alienating for people with disabilities because of the belief that the researcher is the purveyor of specialist knowledge and skills who should be in total control of the research process. In seeking to transform the social relations of research production, Oliver and others note the need for a research paradigm which is not only participatory but, more significantly, emancipatory, a paradigm which should seek to demystify the ideological structures within which (oppressive) power relations are located and create the conditions for processes of self definition, involvement and the taking of political power by people with disabilities (Oliver, 1992, p. 110). Oliver, quoting Giroux (1992), notes how researchers need to:

> understand how subjectivities are produced and regulated through historically produced social forms and how these forms carry and embody particular interests. At the core of the position is the need to develop modes of enquiry that not only investigate how experience is shaped, lived and endured within particular social forms . . . but also how certain apparatuses of power produce forms of knowledge that legitimate a particular kind of truth and way of life' (quoted in Oliver, 1992, p. 110).

This, then, requires, as a first stage, a breaking down of the conventional model of the researcher as 'an expert system' and the development of research methods and procedures which generate the contexts within which participative research evolves. As Zarb (1992) notes, 'historically, disabled people and their representative organizations have been denied the opportunity to influence the agenda for disability research, let alone take control of it' (p. 135). This, then, is a call for a transformation in both the procedural and power relations of social research. Indeed, I concur with Zarb that it is insufficient to involve people with disabilities solely in the early stages of the research process (pre-piloting and piloting), but that there must be involvement in such a way that people with disabilities can frame and elaborate on the research questions throughout the duration of a research project. Very few research projects do this, although there are exceptions.[3]

Indeed, the possibilities for plural forms of research and 'knowledge production' are illustrated by the material practices of feminism and, to a lesser extent, the gay rights movement, both of which have contested conventional forms of knowledge with some success. For instance, the feminist movement has generated new knowledge about the constitution and construction of gender roles, while the emergence of a proactive gay rights movement in the 1970s has transformed the dominant discourses concerning subjects like AIDS towards a more progressive, self referential, positionality (see Sontag, 1991). In particular, such episodes illustrate the Foucauldian conception that power does not emanate from a single centre, like the state, but from a competing plurality of sources, a notion which is, in itself, some justification for the development of a plurality of policy making approaches. In turn, it seems clear that, in identifying what the sources of such an agenda might be, it is vital to avoid conceptualizing the relationship between the social sciences, the state and society as necessarily one whereby the formal institutional apparatus of governance dictates the production and dissemination of knowledge.

Moreover, Stanley and Wise (1993) and Oliver (1992) consider the role of research in facilitating empowerment for people, and, as Oliver notes, the task of emancipatory research is to develop its own understanding of the lived experiences of the research subjects, not to objectify or homogenize the complex social relations underpinning the daily lives of, for example, people with disabilities. In the context of the social relations of research production, Oliver (1992) notes that a combination of approaches is required to provide the means for the empowerment of disabled people, including a (re) description of experience (by people with disabilities themselves) in the face of academic and policy makers who abstract and distort 'local' experiences; a re-definition of the problem, or a concern with policy substance; a challenge to the methodologies of the dominant research paradigms; and the evaluation of the resultant policy programmes by disabled people. There are, however, some difficulties with this overall position, in particular defining what constitutes self knowledge and the extent to which any self knowledge being identified by the individual and/or group can possibly exist outside of wider societal ideologies (or systems of knowledge production) and value systems. Indeed, it seems clear that this is an impossibility and that any empowerment based on people with disabilities taking 'for themselves' is critically dependent on the transformation of power structures and of the means and mechanisms of knowledge production.

In the context of planning for accessible environments, a first stage in the 'appropriation' of knowledge production by people with disabilities is, in part, being addressed by organizations like the RTPI. Planning schools, for instance, are having to demonstrate to RTPI validation committees that trainee planners are being taught about access and disability in the context of equal opportunity issues, while Continuing Professional Development (CPD) for practising planners provides courses raising issues about planning for people with disabilities. However, disability and access still tend to be presented as 'a special issue' and in the planning schools there is little or no interaction between the students and people with disabilities in the wider community. Moreover, attendance at CPD courses is not compulsory so there is little or no guarantee that planners are being trained in access issues. Planning research is also characterized by the relative absence of people with disabilities in the design and implementation of research projects, partly because of their under-representation within academic communities but also because such communities, as Oliver (1992) argues, are resistant to 'opening-up' their research agendas to the specific 'objects' of their research.

This is also reinforced by the funding councils which, while trying to encourage collaborative links between academics and the wider 'policy' community, prevent the latter from directly applying for research funds because they are not recognized as a bona fide research establishment. In turn, this reinforces the pattern of research dependence, or where the 'research initiative' is placed, institutionally, in the control of the 'mainstream' institutions. Moreover, there is a tendency for research about disability to examine people with disabilities and not, as Oliver (1992) has powerfully argued, 'ablebodied society'. In particular, research which seeks to empower needs to follow the observations of Oliver (1992) who concludes that 'it is not a case of educating disabled and ablebodied people for integration, but of fighting institutional disablism; it is not disability relations which should be the field for study but disablism' (p. 112).

CONCLUSIONS

Ultimately, the ableist values and practices of society are deeply embedded, yet a first move for people with disabilities is to assert a 'politics of difference' or one where different types of disabled people seek to emphasize the specificities of themselves, of their needs, affective desires and their particular ways of being. There is no need to be either apologetic or defensive about this because the only alternative is the (re) absorption of the coercive, universalistic, values of the mainstream which seek to deny the vitality of 'other' experiences, other lifestyles. For people with disabilities, however, there are clear dilemmas and contradictions which are, in part, being confronted, not the least of which is, as Barton (1993) argues, the need for urgent debates about issues relating to citizenship and the failure of society to provide the means for all its members to participate. For instance, the absence of adapted public transportation is raised as an issue time and again by people with disabilities in the UK as one of the prime problems in preventing them getting to meetings or going to places where they might be able to influence debates about their wider civil rights and/or liberties. Moreover, there are the problems of developing a collective consciousness between different types of peoples with disabilities and of trying to utilize the diversity of their physiological and environmental experiences as a basis for a broader political movement.

At an international level there is also a range of significant questions about the politics of disablement and social movements. Indeed, it would be problematical to suggest that only western welfare states have pursued strategies of 'normalization' or where a universalistic ideal of the body has been propagated. Indeed, as Young (1990) has argued, state socialist societies have long pursued universalistic ideals, and some of the more militant disabled groups have sprung up in places like Nicaragua and Cuba which, outwardly, have presented themselves as sensitized to the needs of minority and/or oppressed groups. Yet our knowledge of 'states of disability' beyond places like the USA, the UK, Sweden, and one or two other western European countries, is limited, although the phenomenon of physical disability is much more widespread in the 'developing' world. Yet it would be problematical to extend many of the concepts and categories of analysis deployed in this book to the broader global broadcloth, because experiences elsewhere, outside of western Europe, are so diverse and different. This, however, has not prevented the World Health Organization from deploying a universal series of definitions of impairment and disability, so reinforcing the universal desire to delimit, truncate and reduce the essence of human relations to all of a type. Disability, however, is a fluid, transformative, and transforming, state of being, which, ultimately, is neither reducible to a 'type' nor something to be understood solely as an objectifiable phenomenon.

NOTES

1. There is, however, a paradox associated with the pursuit of equal opportunities. While equal opportunities are premised on a critique of the unequal, oppressive, and discriminatory practices of society, those that support policies to 'equalize' opportunity are themselves propagating the idea that discrimination be brought to bear as a way of overturning the unequal opportunities available to people with disabilities. Fullinwider (1980), in the context of the labour market, highlights the difficulty, in that 'if we do not use preferential hiring, we permit discrimination to exist. But preferential hiring is also discrimination. Thus, if we use preferential hiring, we also permit discrimination to exist. The dilemma is that whatever we do, we permit discrimination' (p. 194). Yet, this assumes the moral primacy of policies of non discrimination while arguing that policies which apply the same formal rules to everyone are both, in Young's (1990) terms, 'necessary and sufficient for social justice' (p. 195).

2. In addressing the concerns of Oliver, and others, there are examples of research projects which seem to be breaking down the social relations of research production. For instance, Imrie and McCaskie (1995) have received funding from the Economic and Social Research Council to conduct a fifteen month project looking at the role of access groups in influencing accessibility in the built environment. The design and implementation of the research is occurring through a 'Multi-Interest Discussion Forum' (MIDF), a focus group which was set up by one of the joint applicants, the

Centre For Accessible Environments (CAE). The MIDF comprises twelve people with disabilities from around the UK and it is a participatory mechanism involving the forum in all stages and aspects of the research process.

FURTHER READING

I cannot recommend highly enough the wonderful text by Iris Young (1990) which provides one of the most penetrative accounts of the range of themes which are outlined in this chapter. It is one of the most important texts to have been written on themes of social justice, discrimination and forms of social oppression and will repay a careful reading.

Appendix
A Postal Survey of Disability and Access Issues

SECTION ONE: GENERAL BACKGROUND

(1.1) Name and Position of Respondent ....

(1.2) Name of Local Authority ....

(1.3) Has the Local Authority (the council) formally adopted a policy on access for disabled people? (please circle the relevant category)

YES   NO

IF YES, please answer questions (1.4) and (1.5), otherwise go to question (1.6).

(1.4) When was it adopted

(1.5) What are its main provisions? ....
....
....
....
....

(1.6) Do you have a working definition of disability which is used to inform the development and implementation of access policies? (please circle the relevant category).

YES   NO

IF YES, can you say what it is ....
....
....
....
....
....

SECTION TWO: THE ACCESS OFFICER

(2.1) Is the access officer employed full-time or part-time on access duties? (please circle the relevant category).

Full-time   Part-time

IF PART-TIME, please answer questions (2.2), (2.3), and (2.4) otherwise, go to question (2.5).

(2.2) What are the main duties and official job title of the designated AO (other than the access duties)? ....
....
....
....

(2.3) What proportion (%) of time does the AO spend on access duties? ....
....

(2.4) Why has the Authority not appointed a full-time access officer?..............
....
....
....

(2.5) When was the access officer's job originally created? ....

(2.6) What are the main duties of the access officer?......................................
....
....
....

(2.7) What are the main problems facing the access officer in securing access gains for disabled people? (please tick in the relevant spaces)

	A significant problem	A problem	Not a significant problem	No problem
Lack of time to represent access issues				
Lack of finance and resources				
Indifference amongst planning officers				

Lack of support from
Councillors

Lack of powers to
enforce access
policies

Resistance from
private developers

Other reasons (please specify, or amplify on the points above) ....
....
....
....
....

SECTION THREE: ACCESS AND THE LAND USE PLANNING SYSTEM

(3.1) Which external organizations, if any, do you consult, for guidelines on
access issues? (please tick beside the relevant category).

YES   NO

The Local Access group
Department of the Environment
Access Committee for England
Others (please specify)...
....
....
....

(3.2) Does the authority ever attach planning conditions to a planning
permission decision concerning access for disabled people? (please circle the
relevant category).

YES   NO

IF YES, please answer question (3.3); IF NO, go to question (3.4)

(3.3) Can you indicate what proportion of planning applications have
planning conditions attached to them concerning access for the disabled?
(please tick beside the relevant category, note, if it is difficult to estimate this
please leave unanswered).

All of them
50% to 99% of them
25% to 49% of them
11% to 24% of them
Less than 10%

(3.4) To your knowledge, has a planning application ever been refused on access grounds? (please tick beside the relevant category).

YES   NO

IF YES, can you give details...

....

....

....

....

IF NO, can you say why refusals have never occurred ....

....

....

....

....

(3.5) Does your authority include any statements on access policy in its current statutory plans? (please tick beside the relevant category).

YES   NO

IF YES, can you indicate what references are made

....

....

....

....

IF NO, can you say why there is no reference to access issues ....

....

....

....

....

(3.6). Can you indicate how helpful the building regulations are in providing you with a relevant framework for securing a more accessible environment (please tick beside the relevant category).

Extremely helpful
Very helpful
Helpful
Unhelpful
Very unhelpful
Extremely unhelpful

Can you give reasons for your response? ....

....

....

....

....
....
....

SECTION FOUR: ACCESS AND DEVELOPMENT

(4.1) In your experience, are private sector applicants generally (please place a tick beside the relevant category).

> very willing to build accessible buildings
> willing to build accessible buildings
> need some persuasion but will accede to your requests
> unwilling to build accessible buildings
> very unwilling to build accessible buildings

Can you give me some reasons for your response? ....
....
....
....
....
....
....

(4.2) In your experience, how responsive is your local authority towards the development of accessible public buildings (i.e.. its own public spaces)? (please tick beside the relevant category).

> Extremely responsive
> Very responsive
> Responsive
> Not responsive
> Extremely unresponsive

Can you give me some reasons for your response? ....
....
....
....
....
....
....

SECTION FIVE: ACCESS GROUPS.

(5.1) Is there a local access group? (please tick beside the relevant category)

YES   NO

IF YES, please answer questions (5.2) and (5.3)

(5.2 Does the access group have any of the following involvement in access issues (please tick beside the relevant categories).

YES   NO

a regular inspection of planning applications
regular meetings with the access officer
representation on the Planning Committee
active councillor support

Any other involvement (please specify) ....
....
....
....
....
....
....
....
....

(5.3) In your opinion, do local access groups, or other disability organiz-ations, have (please tick beside the relevant category).

Significant power to influence access policy in the local authority
Some power to influence access policy
Insignificant power to influence access policy
No power at all to influence access policy

Can you give me some reasons for your response? ....
....
....
....
....
....
....
....

THANK YOU FOR FILLING IN THIS QUESTIONNAIRE. CAN YOU
PLEASE NOW PLACE IT IN THE PRE-PAID FOR ENVELOPE AND POST
BACK TO ROYAL HOLLOWAY. ALL RESPONSES WILL REMAIN
CONFIDENTIAL.

References

Abberley, P. (1993) Disabled people and 'normality', in Swain, J., Finkelstein, V., French, S., and Oliver, M. (eds.) *Disabling Barriers - Enabling Environments*, Open University, Milton Keynes, pp. 107-15.

Access Committee for England (1994a) 'PPG1 and 3 fail to help disabled, DoE told by Access Committee', *Planning Week*, 26 May.

Access Committee for England (1994b) *List of Access Officers in England*, ACE, London.

Access Committee for England (1994c) *National Research Project on Local Access Groups, Summary*, ACE, London.

Access Committee for England (1995) *Working Together for Access*, ACE, London.

Adams, C. (1994) *Urban Planning and the Development Process*, UCL Press, London.

Allberry, A. (1992) Letter to David Rose, Director Public Affairs, Royal Town Planning Institute, 4 August.

Ambrose, P. (1994) *Urban Process and Power*, Routledge, London.

Anderson, B. (1990) Statement, Leeds Forum of Disabled People and Equal Opportunities Unit, Leeds, August.

Anderson, E. and Clarke, L. (1982) *Disability and Adolescence*, Methuen, London.

Barker, R. (1977) *Adjustment to Physical Handicap and Illness: A Survey of the Social Psychology of Physique and Disability*, Milwood, New York.

Barnes, C. and Oliver, M. (1995) Disability rights: rhetoric and reality in the UK, *Disability and Society*, Vol. 10, No. 1, pp. 111- 16.

Barnes, C. (1990) *Cabbage Syndrome: The Social Construction of Dependence*, Falmer Press, Basingstoke.

Barnes, C. (1991) *Disabled People in Britain and Discrimination*, Hurst and Company, London.

Barthes, R. (1975) *The Pleasures of the Text*, Hill and Wang, New York.

Barton, L. (ed.) (1989) *Disability and Dependence*, Falmer Press, London.

Barton, L. (1993) The struggle for citizenship: the case of disabled people, *Disability, Handicap, and Society*, Vol. 8, No. 3, pp. 235-48.

Batchelor, R. (1994) *Henry Ford: Mass Production, Modernism, and Design*, Manchester University Press.

Birkenbach, J. (1993) *Physical Disability and Social Policy*, University of Toronto Press.

Blackwell, M., Breston, M., Mayerson, A., and Bailey, S. (1988) Smashing icons: disabled women and the disability and women's movement, in Fine, M. and Asch, A. (1988) *Women with Disabilities*, Temple University Press, Philadelphia, pp. 306-32.

Blaxter, M. (1980) *The Meaning of Disability*, Heinemann, London.

Blomley, N. (1994) Mobility, empowerment, and the rights revolution, *Political Geography*, Vol. 13, No. 5, pp. 407–22.

Blytheway, B. and Johnson, J. (1990) On defining ageism, *Critical Social Policy*, Vol. 11, No. 3, pp. 22-39.

Bordo, S. (1995) *Unbearable Weight: Feminism, Western Culture, and the Body*, University of California Press.

Borsay, A. (1986) *Disabled People in the Community*, Bedford Square Press, London.

Bowe, F. (1978) *Handicapping America*, Harper and Row, New York.

Bradshaw, S. (1995) Comment, *The Guardian*, 13 January, p. 7.

Brooks, S. and Gagnon, A. (eds.) (1991) *Social Scientists, Policy, and the State*, Praeger, New York.

Bruton, M. and Nicholson, D. (1987) *Local Planning in Practice*, Hutchinson, London.

Burgdorf, M. and Burgdorf, R. (1975) A history of unequal treatment: the qualifications of handicapped persons as a suspect class under the equal protection clause, *Santa Clara Lawyer*, Vol. 15, pp. 855-910.

Butler, R. (1995) Visual disability and access, presentation to the session on health and embodiment, Association of American Geographers Conference, Chicago, 12–16 March.

Callinicos, A. (1990) Reactionary postmodernism, in Boyne, R. and Rattansi, A. (eds.), *Postmodernism and Society*, Macmillan, London, pp. 97-118.

Carver, V. and Rhodda, M. (1978) *Disability and the Environment*, Paul Elete, London.

Cohen, J. and Roberts, J. (1983) *On Democracy*, Penguin, New York.

Colomina, B. (ed.) (1992) *Sexuality and Space*, Princeton Architectural Press, New York.

Committee on Restrictions Against Disabled People (CORAD) (1982) *Report*, HMSO, London.

Cooke, P. (1988) Modernity, postmodernity and the city, *Theory, Culture and Society*, Vol. 5, pp. 475-92.

Cotterell, R. (1992) *The Sociology of Law: An Introduction*, Butterworths, London.

Crawford, M. (1992) Can architects be socially responsible?, in Ghirardo, D. (ed.), op. cit, pp. 27–45

Cullingworth, J. and Nadin, V. (1995) *Town and Country Planning in Britain*, Routledge, London.

Curry, D. (1993) Letter to David Rose, Director, Public Affairs, Royal Town Planning Institute, 9 June.

Dalley, G. (ed.) (1991) *Disability and Social Policy*, Policy Studies Institute, London.

Dalley, G. (1992) Social welfare ideologies and normalisation, in Brown, H. and Smith, H. (eds.) *Normalisation: A Reader for the Nineties*, Routledge, London, pp. 100-11.

Darke, P. (1994) The Elephant Man (David Lynch, EMI Films (1980): An analysis from a disabled perspective, *Disability and Society*, Vol. 9, No. 3, pp. 327-43

Daunt, P. (1991) *Meeting Disability: A European Response*, Cassell, London.

Davies, C. and Lifchez, R. (1987) An open letter to architects, in Lifchez, R. (ed.), *Rethinking Architecture*, University of California Press, pp. 35-50.

Davis, M. (1985) Urban renaissance and the spirit of postmodernism, *New Left Review*, Vol. 151, pp. 106-14.

Day, P. (1985) Access and planning at the local level, Proceedings of a conference on implementing accessibility, London.

De Beauvoir, S. (1943) *L'Invitee*, Gallimard, Paris.

De Beauvoir, S. (1976) *The Second Sex*, Penguin, Harmondsworth.

de Jong, G. (1983) Defining and implementing the independent living concept, in Crewe, N. and Zola, I. (eds.), *Independent Living for Physically Disabled People*, Jossey Bass, London.

Dear, M. (1986) Postmodernism and planning, *Environment and Planning D: Society and Space*, Vol. 4, pp. 367-84.

Dear, M. (1995) The body in context, paper presented to the session on Health and Embodiment at the Association of American Geographers Conference, Chicago, 12–16 March.

Department of Employment (1990) *The Employment of People with Disabilities: a Review of the Legislation*. IFF Research Ltd., London.

Department of the Environment (1984) Circular 22/84, *Memorandum on Structure and Local Plans*, HMSO, London.

Department of the Environment (1985a) *Development Control Policy Note 16*, DoE, London.

Department of the Environment (1985b) *Circular 11/85*, DoE, London.

Department of the Environment (1985c) Cmnd 9571, *Lifting the Burden*, White Paper, DoE, HMSO, London.

Department of the Environment (1992a) *Planning Policy Guidance, No. 1*, DoE, London.

Department of the Environment (1992b) *Planning Policy Guidance, No. 3*, DoE, London.

Dickens, P. (1980) Social science and design theory, *Environment and Planning B: Planning and Design*, Vol. 6, pp. 105–17.

Dickenson, M. (1977) Rehabilitating the traumatically disabled adult, *Social Work Today*, Vol. 8, No. 28, p. 12.

Disability Unit (1994) *Summary of a Consultation on Government Measures to Tackle Discrimination Against Disabled People*, DU, The Adelphi, 1-11 John Adam Street, London, WC2N 6HT.

Dougan, G. (1994) Interview, conducted by R. Imrie, 19 March, Washington D.C., USA.

DPTAC (1989) *Public Transport and the Missing Six Millions: What can be Learned*, Disabled Persons Transport Advisory Committee, London.

Drake, R. (1994) The exclusion of disabled people from positions of power in British voluntary organisations, *Disability and Society*, Vol. 9, No. 4, pp. 461–86.

Dreyfuss, H. (1955) *Designing for People*, Simon and Schuster, New York.

Driedger, D. (1989) *The Last Civil Rights Movement: Disabled People's International*, Hurst and Company, London.

Emener, W. (1987) An empowerment philosophy for rehabilitation in the 20th century, *Journal of Rehabilitation*, Vol. 57, p. 7.

Equal Opportunities Division (1993) *Equality in the Civil Service*, Civil Service, London.

Farber, B. (1968) *Mental Retardation: Its Social Context and Social Consequences*, Houghton Mifflin, Boston.

Fine, M. and Asch, A. (1988a) *Women with Disabilities*, Temple University Press, Philadelphia.

Fine, M. and Asch, A. (1988b) Disability beyond stigma: social interaction, discrimination, and activism, *Journal of Social Issues*, Vol. 44, No. 1, pp. 3–21.

Finkelstein, V. (1982) *Attitudes and Disabled People: Issues for Discussion*, World Rehabilitation Fund, New York.

Finkelstein, V. (1988) *Changes in Thinking about Disability*, unpublished paper.

Finkelstein, V. (1993) Workbook 1: being disabled, written for *K665, The Disabling Society*, Open University, Milton Keynes.

Fisher, S. (1973) *Body Consciousness: You Are What You Feel*, Prentice Hall, New Jersey.

Foucault, M. (1980) *Power and Knowledge*, Harvester, Brighton.

Frampton, K. (1992) Reflections on the Autonomy of Architecture: A Critique of Contemporary Production, in Ghirardo, D. (ed.), op. cit., pp. 17–26.

Freeman, A. (1982) Antidiscrimination law: a critical review, in Karys, D. (ed.) *The Politics of Law: A Progressive Critique*, Pantheon, New York.

French, S. (1993) Disability, impairment, or something in between, in Swain, J., Finkelstein, V., French, S., and Oliver, M. (eds.), op. cit., pp. 17–25.

Fullinwider, R. (1980) *The Reverse Discrimination Controversy*, Rowman and Allanheld, Totowa, New Jersey.

Gardiner, S. (1974) *Le Corbusier*, Fontana, London.

Garety, P. and Toms, R. (1990) Collected and neglected – are Oxford hostels for the homeless filling up with disabled psychiatric patients, *British Journal of Psychiatry*, Vol. 157, No. 8, pp. 269–72.

Gertner, V. (1994) Interview conducted by R. Imrie, Berkeley, California, 3 April .

Ghirardo, D. (ed.) (1991) *Out of Site: A Social Criticism of Architecture*, Bay Press, Seattle.

Giddens, A. (1991) *Modernity and Self Identity: Self and Security in the Late Modern Age*, Polity Press, Cambridge.

Gilderbloom, J. and Rosentraub, M. (1990) Creating the accessible city: proposals for providing housing and transportation for low income, elderly, and disabled people, *American Journal of Economics and Sociology*, Vol. 49, No. 3, pp. 271–82.

Gilroy, R. (1993) *Good Practice in Equal Opportunities*, Avebury, Aldershot.

Giroux, H. (1992) *Border Crossings: Cultural Workers and the Politics of Education*, Routledge, London.

Gleeson, B. (1995) *A Geography for Disabled People?*, unpublished manuscript.

Goffman, E. (1963) *Stigma: Notes on the Management of Spoiled Identity*, Simon and Schuster, New York.

Golledge, R. (1991) Special populations in contemporary urban areas, in Hart, J. (ed.), *Our Changing Cities*, Johns Hopkins University Press, Baltimore, pp. 146–69.

Golledge, R. (1993) Geography and the disabled: a survey with special reference to vision impaired and blind populations, *Transactions of the Institute of British Geographers*, New Series, Vol. 18, pp. 63–85.

Gooding, C. (1994) *Disabling Laws, Enabling Acts: Disability Rights in Britain and America*, Pluto Press, London.

Goodwin, S. (1995) Inside Parliament, *The Independent*, 25 January, p. 5.

Graham, J. (1990) Theory and essentialism in Marxist geography, *Antipode*, Vol. 22, pp. 53–66.

Gramsci, A. (1971) *The Prison Notebooks*, Lawrence and Wishart, London.

Grosz, E. (1992) Bodies-cities, in Colomina, B. (ed.) (1992) *Sexuality and Space*, Princeton Architectural Press, New York, pp. 241–54.

Habermas, J. (1975) *Legitimation Crisis*, Beacon Press, Boston.

Habermas, J. (1987) *The Theory of Communicative Competence, Volume 2: Lifeworld and System*, Beacon, Boston.

Hahn, H. (1986a) Disability and the urban environment: a perspective on Los Angeles, *Environment and Planning D: Society and Space*, Vol. 4, pp. 273–88.

Hahn, H. (1986b) Public support for rehabilitation programs: the analysis of US disability policy, *Disability, Handicap, and Society*, Vol. 1, No. 2, pp. 121–38.

Hahn, H. (1988) The politics of physical differences: disability and discrimination, *Journal of Social Issues*, Vol. 44, No. 1, pp. 39–47.

Hall, E. (1995) Contested (dis) abled identities in the urban labour market, paper presented at the 10th Urban Change and Conflict Conference, Royal Holloway, University of London, Egham, Surrey TW20 0EX, 5–7 September.

Hall, P. (1984) *Urban and Regional Planning*, Penguin, Harmondsworth.

Hammersley, M. (1992) On feminist methodology, *Sociology*, Vol. 26, pp. 187–206.

Hari, M. (1975) The idea of learning in conductive pedagogy, in Akos, K. (ed.), *Scientific Studies on Conductive Pedagogy*, Institute for Motor Disorders, Budapest.

Harris, A., Cox, E., and Smith, R. (1971) *Handicapped and Impaired in Great Britain*, HMSO, London.

Harvey, D. (1973) *Social Justice and the City*, Edward Arnold, London.

Harvey, D. (1990) *The Condition of Postmodernity*, Blackwell, Oxford.

Hayden, D. (1981) What would a non sexist city be like: speculations on housing, urban design, and human work, in Stimpson, C., Dixler, E., Nelson, M., and Yatrakis, K., (eds.), *Women and the American City*, University of Chicago Press.

Healey, P. (1992) The reorganisation of state and market, *Urban Studies*, Vol. 29, pp. 411–34.

Healey, P. (1995) Discourses of integration: making frameworks for democratic urban planning, in Healey, P., Cameron, S., Davoudi, S., Graham, S., and Madani-Pour, A. (eds.) *Managing Cities: The New Urban Context*, Wiley, London, pp. 251–72.

Healey, P., McNamara, P., Elson, M., and Doak, A. (1988) *Land Use Planning and the Mediation of Urban Change*, Cambridge University Press.

Hellman, C. (1984) *Culture, Health, and Illness*, John Wright and Sons, Bristol.

Heron, J. (1981) Experimental research methods, in Reason, P. and Rowan, J. (eds.) *Human Inquiry*, Wiley, New York, pp. 153–66.

Hevey, D. (1991) From self love to the picket line, in Lees, S. (ed.) *Disability Arts and Culture Papers*, Shape Publications, London.

Higgens, P. (1992) *Making Disability*, Charles Thomas, Springfield, Illinois.

Hillier, J. (1993) Using NVQs flexibly, *The NVQ Monitor*, Autumn, NCVQ, London.

HMSO (1989) *Progress on Cities*, HMSO, London.

Hobsbawn, E. (1968) *Industry and Empire*, Penguin, Harmondsworth.

Hooks, B. (1984) *Feminist Theory: From Margin to Center*, South End Press, Boston.

House Builders Federation (1995) The application of building regulations to help disabled people in new dwellings in England and Wales, unpublished paper, HBF, London.

Howard, E. (1898) *Tomorrow: A Peaceful Path to Real Reform*, Swan Sonnenschein, London.

Howard, E. (1902) *Garden Cities of Tomorrow*, Swan Sonnenschein, London.

Imrie, R. (1996a) Transforming the social relations of research production in urban policy evaluation, *Environment and Planning A*, (in press).

Imrie, R. (1996b) Ableist geographies, disablist spaces: towards a reconstruction of Golledge's geography and the disabled, *Transactions of the Institute of British Geographers, New Series*, Vol. 21, No. 2 (in press).

Imrie, R. (1996c) Equity, social justice, and planning for access and disabled people: an international perspective, *International Planning Studies*, Vol. 1, No. 1, pp. 17–34.

Imrie, R. and McCaskie, K. (1995) Disability, access, and the built environment, *Access By Design*, Centre for Accessible Environments, London.

Imrie, R. and Wells, P. (1992) Creating the barrier free environment: the planning system and access in South Wales, *Town and Country Planning*, November, pp. 278–80.

Imrie, R. and Wells, P. (1993) Disablism, planning, and the built environment, *Environment and Planning C: Government and Policy*, Vol. 11, No. 2, pp. 213–31.

Independent Living Consumer Survey (1991) *Independent Living*, May–June issue, p. 3.

Innovations (1992) 10 facts about housing and mobility problems, *Innovations in Social Housing*, Vol. 3, p. 5.

Jacobs, J. (1961) *The Death and Life of Great American Cities*, Random House, New York.

Jameson, F. (1991) *Postmodernism, or the Cultural Logic of Late Capitalism*, Verso, London.

Jencks, C. (1987) *Le Corbusier and the Tragic View of Architecture*, Penguin, Harmondsworth.

Jenkins, B. and Gray, A. (1991) Evaluation and the consumer: the UK experience, in Mayne, J., Bemelmas-Videc, M., Hudson, J., and Connor, R. (eds.) *Advancing Public Policy Evaluation*, North Holland Press, Amsterdam, pp. 285–99.

Johnson, M. (1988) Disability rights movement – overcoming the social barriers, *Nation*, Vol. 246, No. 14, p. 489.

Johnson, M. (1994) Sticks and stones: the language of disability, in Nelson, J. (ed.) *The Disabled, the Media, and the Information Age*, pp. 25–43.

Jordanova, L. (1989) *Sexual Visions*, Harvester Wheatsheaf, New York.

Kasprzyk, D. (1983) Psychological factors associated with response to hypertension or spinal cord injury: an investigation of coping with chronic illness or disability, *Dissertation Abstracts International*, Vol. 44, No. 4B, p. 1279.

Kettle, M. (1979) *Disabled People and their Employment: A Review of Research into the Performance of Disabled People at Work*, Association of Disabled Professionals, Surrey.

King, A. (ed.) (1984) *Buildings and Society*, Routledge and Kegan Paul, London.

Knesl, J. (1984) The powers of architecture, *Environment and Planning D: Society and Space*, Vol. 2, pp. 3–22.

Knox, P. (1987) The social production of the built environment – architects, architecture, and the post modern city, *Progress in Human Geography*, Vol. 11, No. 3, pp. 354–78.

Kristeva, J. (1982) *Powers of Horror: An Essay in Abjection*, Colombia University Press, New York.

Lacan, J. (1977) *Ecrits: A Selection*, Tavistock, London.

Laclau, E. and Mouffe, C. (1985) *Hegemony and Socialist Strategy: Towards a Radical Democratic Politics*, Verso, London.

Lambert, S. (1993) *Form Follows Function: Design in the 20th Century*, Victoria and Albert Museum, London.

Lane, H. (1995) Constructions of deafness, *Disability and Society*, Vol. 10, No. 2, pp. 171–90.

Langton-Lockton, S. (1994) Review of the European manual for the accessible environment, revision concept 2, *Access by Design*, Vol. 64, p. 21.

Lash, S. and Urry, J. (1987) *The End of Organised Capitalism*, Polity Press, Cambridge.

Lather, P. (1991) *Getting Smart: Feminist Research and Pedagogy With/In the Postmodern*, Routledge, London.

Laune, L. (1993) *Building Our Lives: Housing, Independent Living, and Disabled People*, Conference Report, Shelter, London.

Laurance, J. (1995a) Minister pledges rights for disabled – in 15 years' time, *The Independent*, 13 January, p. 4.

Laws, G. (1994a) Oppression, knowledge, and the built environment, *Political Geography*, Vol. 13, No. 1, pp. 7–32.

Laws, G. (1994b) Ageing, contested meanings, and the built environment, *Environment and Planning A*, Vol. 26, No. 11, pp. 1787–1802.

Le Corbusier (1927) *Towards a New Architecture*, Architectural Press, London, translated by Etchells, F.

Leach, B. (1989) Disabled people and the implementation of local authorities equal opportunities policies, *Public Administration*, Vol. 67, pp. 65–77.

Lennon, K. and Whitford, M. (1994) *Knowing the Difference: Feminist Perspectives in Epistemology*, Routledge, London.

Leonard, B. (1978) *Impaired View: Television Portrayals of Handicapped People*, PhD thesis, Boston University, USA.

Liachowitz, C. (1988) *Disability as a Social Construct*, University of Pennsylvania, USA.

Lifchez, R. and Winslow, B. (1979) *Design for Independent Living*, University of California Press.

Little, J. (1994) *Gender, Planning, and the Policy Process*, Pergamon, Oxford.

Little, J., Peake, L., and Richardson, P. (eds.) (1988) *Women in Cities: Gender and the Urban Environment*, Macmillan, London.

MacDonald, R. (1995) Disability and planning policy guidance, paper presented to the Access Sub Committee, Oxford City Council, 7 March.

Marcuse, H. (1964) *One Dimensional Man*, Beacon, Boston.

Marshall, T. (1950) *Citizenship and Social Class*, Cambridge University Press.

Martin, J. and Elliot, D. (1992) Creating an overall measure of severity of disability for the OPCS disability survey, *Journal of the Royal Statistical Society Series A - Statistics in Society*, Vol. 155, No. 1, pp. 121–40.

Marx, K. (1969) *Grundrisse*, Pelican, London.

Matrix (1984) *Making Space: Women and the Man-Made Environment*, Pluto Press, London.

McAllister, D. (1980) *Evaluation in Environmental Planning*, MIT Press, London.

McGlynn, S. and Murrain, P. (1994) The politics of urban design, *Planning Practice and Research*, Vol. 9, No. 3, pp. 311–320.

Millerick, M. and Bate, R. (1991) I can, planners can: planning for disabled people, *Housing and Planning Review*, June/July, pp. 12–13.

Mitchell, D. (1995) The end of public space? People's park, definitions of the public, and democracy, *Annals of the Association of American Geographers*, Vol. 85, No. 1, pp. 108–33.

Moore, C. and Bloomer, K. (1977) *Body, Memory, and Architecture*, Yale University Press, New Haven.

Morris, J. (1991) *Pride against Prejudice: Transforming Attitudes to Disability*, Women's Press, London.

Morris, J. (1992) 'Us' and 'them'? Feminist research, community care, and disability, *Critical Social Policy*, Vol. 11, No. 3, pp. 22–39.

Morris, J. (1993) *Independent Lives*, Macmillan, London.

Murphy, R. (1987) *The Body Silent*, Phoenix House, London.

Myette, P. (1994) Interview conducted by R. Imrie, 18 April, Boston, USA.

Nelson, J. (1994) Broken images: portrayals of those with disabilities in American media, in Nelson, J. (ed.) *The Disabled, the Media, and the Information Age*, Greenwood Press, Westport, pp. 1–17.

Nevins, D. (1981) From eclecticism to doubt, *Heresies*, Vol. 11, No. 3, pp. 71–2.

Nichols, J. (1995) Interview conducted by R. Imrie, 28 July.

O'Neil, J. (1995) *The Poverty of Postmodernism*, Routledge, London.

Office of Population Censuses and Surveys (1971) *The Prevalence of Disability in Great Britain*, HMSO, London.

Office of Population Censuses and Surveys (1987) *The Prevalence of Disability in Great Britain, Report 1*, HMSO, London.

Office of Population Censuses and Surveys (1989) *Surveys of Disability in Great Britain*, HMSO, London.

Oliver, M. (1984) The politics of disability, *Critical Social Policy*, Vol. 4, No. 2, pp. 21–32.

Oliver, M. (1990) *The Politics of Disablement*, Macmillan, London.

Oliver, M. (1992) Changing the social relations of research production, *Disability, Handicap, and Society*, Vol. 7, No. 2, pp. 101–14.

Pagel, M. (1988) *On Our Own Behalf*, Greater Manchester Coalition of Disabled People.

Pawley, M. (1973) *The Private Future: Causes and Consequences of Community Collapse in the West*, Thames & Hudson, London.

Pawley, M. (1983) The defence of modern architecture, *RIBA Transactions 2*, pp. 50–5.

Pile, S. and Thrift, N. (eds.) (1995) *Mapping the Subject*, Routledge, London.

Porphyrios, D. (1985) On critical history, in Ockman, J. (ed.) *Architecture, Criticism, Ideology*, Princeton Architectural Press, New York, pp. 16–21.

Prak, N. (1984) *Architects: the Noted and the Ignored*, Wiley, Chicester.

Priestly, M. (1995) Commonality and difference in the movement: an association of blind Asians in Leeds, *Disability and Society*, Vol. 10, No. 2, pp. 157–70.

RADAR (1993) *Disability and Discrimination in Employment*, RADAR, London.

Rawls, J. (1971) *A Theory of Justice*, Harvard University Press, Boston.

Rees, G., Williamson, H. and Winckler, V. (1989) *The New Vocationalism, Further Education and Local Labour Markets*, manuscript available from G. Rees, UWCC, Department of Sociology, Park Place, Cardiff.

Reeves, D. (1995) Mobility and accessible housing – tokenism to positivism. Unpublished paper, available from the author, Centre for Planning, University of Strathclyde, Richmond Street, Glasgow, G1.

Reisen, L. (1990) Autonomous living and household management of disabled elderly by support of new technologies, *Zeitschrift für Georontologie*, Vol. 23, No. 1, pp. 3–11.

Rose, G. (1990) The struggle for political democracy: emancipation, gender, and geography, *Society and Space*, Vol. 8, No. 4,, pp. 395–408.

Royal Institute of British Architects (1993) *Examinations in Architecture, Description and Regulations*, RIBA, London.

Royal Town Planning Institute (1988) Access for disabled people, *Practice Advice Note*, No. 3, RTPI, London.

Royal Town Planning Institute (1993) *Access Policies for Local Plans*, RTPI, London.

Royal Town Planning Institute (1995) *Practice Advice Note*, No. 9, RTPI, London.

Ryan, J. and Thomas, F. (1980) *The Politics of Mental Handicap*, Penguin, Harmondsworth.

Rydin, Y. (1993) *The British Planning System: An Introduction*, Macmillan, London.

Salman, J. (1994) Interview conducted by R. Imrie, 28 March.

Savage, M. and Warde, A. (1993) *Urban Sociology, Capitalism, and Modernity*, Macmillan, London.

Sayer, A. (1984) *Method in Social Science*, Hutchinson, London.

Scheer, J. and Groce, N. (1988) Impairment as a human constraint: cross cultural and historical perspectives on variation, *Journal of Social Issues*, Vol. 44, No. 1, pp. 23–37.

Schull, A. (1984) A convenient place to get rid of inconvenient people: the Victorian lunatic asylum, in King, A. (ed.) *Buildings and Society*, Routledge and Kegan Paul, London, pp. 37–60.

Scotch, R. (1988) Disability as the basis for a social movement: advocacy and the politics of definition, *Journal of Social Issues*, Vol. 44, No. 1, pp. 159–72.

Scotch, R. (1989) Politics and policy in the history of the disability rights movement, *Millbank Quarterly*, Vol. 67, No. 52, pp. 380–400.

Scott, V. (1993) *Lessons From America: A Study of the ADA*, Access Committee for England, London.

Sennett, R. (1990) *The Conscience of the Eye: The Design and Social Life of Cities*, Faber and Faber, London.

Sennett, R. (1994) *Flesh and Stone*, Faber and Faber, London.

Shakespeare, T. (1993) Disabled people's self-organisation: a new social movement?, *Disability, Handicap, and Society*, Vol. 8, No. 3, pp. 249–64.

Shakespeare, T. (1994) Cultural representations of disabled people: dustbins for disavowal, *Disability and Society*, Vol. 9, No. 3, pp. 283–300.

Shearer, A. (1981) *Disability: Whose Handicap?*, Blackwell, Oxford.

Showalter, E. (1987) *The Female Malady: Women, Madness, and English Culture, 1830–1980*, Penguin, New York.

Silver Jubilee Committee (1979) Can disabled people go where you go? *Silver Jubilee Access Committee Report*, Department of Health and Social Security, London.

Slee, R. (1993) The politics of integration – new sites for old practices, *Disability and Society*, Vol. 8, No. 4, pp. 351–60.

Smart, C. (1989) *Feminism and the Power of Law*, Routledge, London.

Smith, M. and Judd, D. (1984) American cities: the production of ideology, in Smith, M. (ed.) *Cities in Transformation*, Sage, Berkeley.

Soder, M. (1984) The mentally retarded: ideologies of care and surplus population, in Barton, L. and Tomlinson, S. (eds.) *Special Education and Social Interest*, Croom Helm, Andover.

Sontag, S. (1991) *Illness as Metaphor/AIDS and its Metaphors*, Penguin, London.

Speare, A., Avery, R. and Lawton, L. (1991) Disability, residential mobility, and changes in living arrangements, *Journal of Gerontology*, Vol. 46, No. 3, pp. 133–42.

Stanley, L. and Wise, S. (1993) *Breaking Out Again: Feminist Ontology and Epistemology*, Routledge, London.

Steinberger, P. (1985) *Ideology and the Urban Crisis*, Albany State University Press, New York.

Stone, D. (1984) *The Disabled State*, Macmillan, London.

Sullivan, L. (1947) *Kindergarten Chats and Other Writings*, Wittenborn Shultz, New York.

Summers, D. (1992) Disabled quota not being met, *The Financial Times*, 6 February, p. 3.

Svizos, S. (1992) The limits to integration, in Brown, H. and Smith, H. (eds.) *Normalisation: A Reader for the Nineties*, Routledge, London, pp. 112–33.

Swain, J., Finkelstein, V., French, S., and Oliver, M. (eds.) (1993) *Disabling Barriers – Enabling Environments*, Open University, Milton Keynes

Taylor-Gooby, P. (1994) Postmodernism and social policy: a great leap backwards?, *Journal of Social Policy*, Vol. 23, No. 3, pp. 385–404.

tenBroek, J. (1966) 'The right to live in the world': the disabled in the law of torts, *California Law Review*, Vol. 54, No. 84, pp. 1

Thomas, H. (1992) Disability, politics, and the built environment, *Planning Practice and Research*, Vol. 7, No. 1, pp. 22–5.

Thomas, H. (1995) 'Race', public policy, and planning in Britain, *Planning Perspectives*, Vol. 10, pp. 123–48.

Thornley, A. (1991) *Urban Planning Under Thatcherism: The Challenge of the Market*, Routledge, London.

Tomlinson, S. and Colquhoun, R. (1995) The political economy of special educational needs in Britain, *Disability and Society*, Vol. 10, No. 2, pp. 191–202.

Townsend, P. (1979) *Poverty in the United Kingdom*, Penguin, Harmondsworth.

Trade Union Congress (1993) *Trade Unions and Disabled Members*, TUC, London.

Trent, J. (1994) *Inventing the Feeble Mind: A History of Mental Retardation in the United States*, University of California Press.

US Department of Labor (1948) *The Performance of Physically Impaired Workers in Manufacturing Industries*, US Department of Labor, Washington D.C.

Vaughan (1991) The social bases of conflict between blind people and agents of rehabilitation, *Disability, Handicap and Society*, Vol. 6, No. 3, pp. 203–17

Vujakovic, P. and Mathews, M. (1994) Contorted, folded, torn: environmental values, cartographic representation, and the politics of disability, *Disability and Society*, Vol. 9, No. 3, pp. 359–74.

Wajcman, J. (1991) *Feminism Confronts Technology*, Polity, Oxford.

Walker, A. (1994) Interview conducted by R. Imrie, Architectural Association, London.

Walker, R. and Sayer, A. (1993) *The New Social Economy*, Blackwell, Oxford.

Ward, S. (1994) *Planning and Urban Change*, Paul Chapman, London.

Wates, N. and Knevitt, C. (1987) *Community Architecture: How People are Creating their Own Environments*, Penguin, London.

Watson, S. and Gibson, K. (1995) *Postmodern Cities and Spaces*, Blackwell, Oxford.

Weisman, L. (1992, *Discrimination by Design*, University of Illinois Press.

Welsh Council for the Disabled (1987) *The Implementation of Access Legislation in Local Authorities in Wales*, Access Committee for Wales, Cardiff.

Welsh Council for the Disabled (1990) *Access Officers in Wales*, Access Committee for Wales, Cardiff.

Which? (1989) 'No Entry', October, pp. 498–501.

Williams, F. (1992) Somewhere over the rainbow: universality and diversity in social policy, in Manning, N. and Page, R. (eds.) *Social Policy Review*, Vol. 4, Social Policy Association, Canterbury, pp. 200–19.

Williams, P. (1987) Alchemical notes: reconstructing ideals from deconstructed rights, *Harvard Civil Rights – Civil Liberties Review*, Vol. 22.

Wirth, L. (1936) Urbanism as a way of life, *American Journal of Sociology*, Vol. XLIV, No. 1, pp. 1–24.

Wolf, N. (1990) *The Beauty Myth: How Images of Beauty Are Used Against Women*, Vintage, London.

Wolfe, A. (1977) *The Limits of Legitimacy: Political Contradictions of Contemporary Capitalism*, Free Press, New York.

Wolfe, T. (1981) *From Bauhaus to Our House*, Farar, Straus, Giroux, New York.

Wolfensberger, W. (1975) *The Origin and Nature of our Institutional Models*, Human Policy Press, Syracuse.

Wolfensberger, W. (1983) Social role valorisation: a proposed new term for the principle of normalisation, *Mental Retardation*, Vol. 21, No. 6, pp. 234–39.

Women in Geography Study Group (1984) *Geography and Gender: An Introduction to Feminist Geography*, Hutchinson, London.

Wood, P. (1981) *International Classification of Impairments, Disabilities, and Handicaps*, World Health Organisation, Geneva.

Wright, G. (1991) *The Politics of Design in French Colonial Urbanism*, University of Chicago Press.

York City Council (1993) *York Access Design Guide: Access to Buildings and the Spaces around them for People with Disabilities*, Directorate of Development Services, York.

Young, I. (1990) *Justice and the Politics of Difference*, Princeton University Press, New York.

Zarb, G. (1992) On the road to Damascus: first steps towards changing the relations of disability research production, *Disability, Handicap, and Society*, Vol. 7, No. 2, pp. 125–38.

Index